AF581880

Arthropod-Borne Viruses

Arthropod-Borne Viruses: The Outbreak Edition

Editors

Rebekah C. Kading
Aaron C Brault
J. David Beckham

MDPI • Basel • Beijing • Wuhan • Barcelona • Belgrade • Manchester • Tokyo • Cluj • Tianjin

Editors
Rebekah C. Kading
Colorado State University
USA

Aaron C Brault
Centers for Disease Control and Prevention
USA

J. David Beckham
University of Colorado School of Medicine
USA

Editorial Office
MDPI
St. Alban-Anlage 66
4052 Basel, Switzerland

This is a reprint of articles from the Special Issue published online in the open access journal *Tropical Medicine and Infectious Disease* (ISSN 2414-6366) (available at: https://www.mdpi.com/journal/tropicalmed/special_issues/Arthropod_Borne_Viruses).

For citation purposes, cite each article independently as indicated on the article page online and as indicated below:

LastName, A.A.; LastName, B.B.; LastName, C.C. Article Title. *Journal Name* **Year**, *Article Number*, Page Range.

ISBN 978-3-03943-348-3 (Hbk)
ISBN 978-3-03943-349-0 (PDF)

Contents

About the Editors

Rebekah C. Kading, Ph.D.: Dr. Kading obtained her BS in Entomology/Wildlife Conservation from the University of Delaware, MS in Entomology from the University of Arkansas, and Ph.D. in Molecular Microbiology and Immunology from the Johns Hopkins Bloomberg School of Public Health. Between 2007 and 2014, Dr. Kading led studies on the ecology of arthropod-borne viruses in Colorado, Uganda, and Guatemala, and the transmission of Rift Valley fever virus (RVFV) by mosquitoes at the CDC Division of Vector-borne Diseases. Her research program is currently focused on RVFV transmission, as well as the ecology of virus circulation among bats.

Aaron C Brault obtained his BS in Zoology from Texas A&M University and Ph.D. from the University of Texas Medical Branch. He then completed post-doctoral training as an American Society of Microbiology fellow at the Division of Vector-Borne Diseases (DVBD), Centers for Disease Control and Prevention in Fort Collins, Colorado, from 2001–2003. Following his ASM fellowship, he accepted a position as an associate professor at the Center for Vector-Borne Diseases and Department of Pathology, Microbiology and Immunology within the School of Veterinary Medicine at the University of California, Davis. After being promoted to associate professor, Dr. Brault returned to DVBD in 2009 as a staff scientist. His laboratory focused on arboviral pathogenesis and molecular genetics, and in 2019, was appointed as the lead for the arboviral diagnostics group within the branch.

J. David Beckham obtained his MD from Baylor College of Medicine, followed by Internal Medicine Residency training at Baylor College of Medicine and Infectious Disease Fellowship Clinical Training at University of Colorado School of Medicine. He then completed post-doctoral training in virology at University of Colorado School of Medicine from 2005 to 2010. Since 2010, Dr. Beckham's laboratory has studied the pathogenesis of flavivirus infections and mechanisms of neuroinflammatory responses to acute viral infections in the central nervous system.

Tropical Medicine and Infectious Disease

Editorial

Global Perspectives on Arbovirus Outbreaks: A 2020 Snapshot

Rebekah C. Kading [1,*], Aaron C. Brault [2] and J. David Beckham [3]

1 Department of Microbiology, Immunology, and Pathology, Colorado State University, Campus Delivery 1690, Fort Collins, CO 80523, USA
2 Division of Vector-borne Diseases, Arboviral Diseases Branch, Centers for Disease Control and Prevention, Fort Collins, CO 80521, USA; zlu5@cdc.gov
3 Departments of Medicine, Neurology, and Immunology & Microbiology, Division of Infectious Diseases, University of Colorado School of Medicine, Rocky Mountain Regional VAMC, 12,700 E. 19th Avenue, B168, Denver, CO 80045, USA; david.beckham@cuanschutz.edu
* Correspondence: rebekah.kading@colostate.edu; Tel.: +1-970-491-7833

Received: 2 September 2020; Accepted: 4 September 2020; Published: 7 September 2020

Keywords: mosquito; tick; emerging infectious diseases; one health; vector-borne diseases

---

When this special issue was first conceived in early 2019, we never anticipated that the publication of this collection of articles would be happening during a pandemic. While this outbreak collection is focused on viruses transmitted by arthropods, its release concurrent with the SARS-CoV-2 pandemic and international health emergency provides an appropriate context in which to draw attention to research focused on other high-consequence, epidemic-causing viruses that may be next to emerge on the global stage.

Arthropod-borne pathogens account more than 17% of infectious diseases, affect millions of people around the world each year, and comprise a significant proportion of emerging human pathogens [1–4]. Dengue, as the most widespread arboviral disease, causes more than 90 million cases and approximately 40,000 deaths per year [3]. The emergence and explosive spread of Zika virus (ZIKV) has recently challenged the global health infrastructure to diagnose and differentiate ZIKV infections from infections with closely related arboviruses and to minimize the catastrophic effects of congenital infection. Additional arboviruses including Rift valley fever, Mayaro, West Nile, chikungunya, and tick-borne encephalitis viruses have captured the attention of the scientific community as emerging public health threats [3,5]. As the world responds to outbreak after outbreak, research on: surveillance and preparedness, virus transmission dynamics, ecology and epidemiology, diagnosis and treatment, public attitudes and practices and existing barriers and challenges to outbreak prevention and control of these and other emerging viruses will be needed to contribute to more effective mitigation strategies and for protecting public health.

This Special Issue comprises a highly diverse group of articles from around the world, which highlight various aspects of arbovirus outbreaks in endemic regions and in areas of introduction. This issue showcases the important work that is being done to mitigate epidemic activity and protect human and animal health from emerging arboviral threats, as well as provides a unique insight into past epidemics. Articles in this issue include coverage of viruses transmitted by ticks as well as mosquitoes, discussions of clinical, epidemiological, and entomological perspectives, and includes emerging arboviruses from every hemisphere. Below is a brief tour of the articles in this collection (Figure 1).

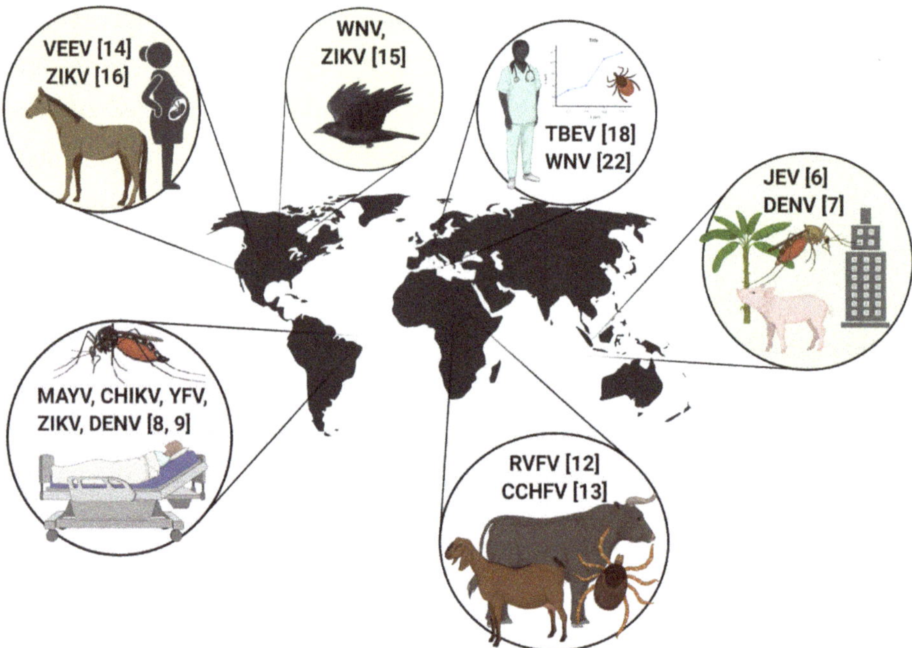

Figure 1. International scope of arbovirus outbreak-related articles included in the "Arthropod-borne Viruses: The Outbreak Edition" special collection. Venezuelan equine encephalitis virus (VEEV), Zika virus (ZIKV), West Nile virus (WNV), tick-borne encephalitis virus (TBEV), Japanese encephalitis virus (JEV), dengue virus (DENV), Rift Valley fever virus (RVFV), Crimean-Congo hemorrhagic fever virus (CCHV), Mayaro virus (MAYV), chikungunya virus (CHIKV), and yellow fever virus (YFV). Numbers indicate the associated reference.

Southeast Asia: The two studies included in this collection from Southeast Asia provide corresponding perspectives on the transfer or "spillover" of zoonotic pathogens from natural transmission cycles into the human population. Young et al. [6] described the blood feeding patterns of mosquitoes in Malaysian Borneo, in an area undergoing significant changes in land cover and land use. They provided novel insights into how the host utilization patterns of some major mosquito vectors in different land use types could influence the spread and spillover of arboviruses, including Japanese encephalitis virus, between natural and epidemic cycles. As a complement to this work, Ab Hamid et al. [7] studied the vertical stratification of *Aedes* mosquitoes responsible for the transmission of dengue virus (DENV) in high-rise buildings in Malaysian cities. They reported mosquito infestations up to 21 stories high, demonstrating a unique risk of arbovirus transmission in urban settings [7].

South America: While screening sera from human patients in Sinop city, Brazil, de Silva Pessosa Vierra et al. [8] documented concurrent circulation of both Mayaro (MAYV) and chikungunya (CHIKV) viruses. While the MAYV genomes detected matched other strains known to be present, the CHIKV genome detected from one patient was of the East/Central/South African (ECSA) genotype. This genotype was distinct from strains already circulating in this area and represented an important discovery, that there was a separate introduction event of CHIKV into Brazil [8]. The simultaneous circulation of *Aedes* mosquito-borne viruses including DENV, CHIKV, ZIKV, and MAYV, and the stress this poses to the human population and health infrastructure in Brazil is discussed in detail by Magalhaes et al. [9].

Africa: This collection covers two hemorrhagic fever viruses endemic to Africa, Rift Valley fever virus (RVFV) and Crimean Congo hemorrhagic fever virus (CCHFV), both of which are listed as "select agents" [10] due to the severity of disease they could cause, as well as potential use as agents of bioterrorism. Rift Valley fever virus is endemic to Africa where it causes large epizootics typically associated with climate patterns and rainfall, and causes significant morbidity and mortality in both humans and animals [11]. This virus has spread to Saudi Arabia, Madagascar, and other island nations in the Indian Ocean (i.e., Mayotte, Comoros), but has not yet emerged in a transoceanic location. Mitigation of a RVFV outbreak carries with it many unique considerations, and would involve mobilization of diverse agencies focused on public health, animals and agriculture, and biosecurity. To this end, Grossi-Soyster and LaBeaud [12] reviewed major considerations for such risk mitigation pertaining to future outbreaks of RVFV, with a special emphasis on vaccines, travelers and tourism. Crimean Congo hemorrhagic fever virus, transmitted by ticks, was originally described from both central Africa as well as the Crimean region of Russia. This virus has alarmingly sustained continual epizootic activity from Africa north through Western Asia in recent years. Sorvillo et al. [13] comprehensively reviewed and discussed a One Health approach to CCHFV prevention and included a special focus on knowledge gaps that are critically important to address.

North America: The United States has experienced invasions of West Nile and Zika viruses in recent decades, with continual threats of endemic re-emerging mosquito-borne viruses such as St. Louis, LaCrosse, Powassan, and Eastern equine encephalitis viruses. In 1971, a transboundary outbreak of Venezuelan equine encephalitis (VEE) virus epidemic strain 1b invaded South Texas. In a letter to the editor, McLean [14] provided a valuable first-hand account of the interagency response to this outbreak, and how this outbreak response remains the sole example of the successful prevention of VEE establishment in the United States. To build on this historical perspective, Kading et al. [15] provided a 30-year analysis on the reactionary response of funding agencies to the emergence and invasion of mosquito-borne viruses in the Americas and how these events have also stimulated the innovative development of traps and augmented surveillance capacity. As the most recent arbovirus introduction to the United States, ZIKV infections have been mostly associated with travelers, however local transmission was documented in areas of Florida and Texas [16,17]. To this end, Hinojosa et al. [16] reported locally-acquired cases of ZIKV in South Texas, particularly as surveillance efforts were increased out of specific concerns of virus infections in pregnant women.

Europe. Tick-borne encephalitis (TBE) is a severe infection of the central nervous system, caused by tick-borne encephalitis virus (TBEV) in the family *Flaviviridae*. Incidence of TBE has increased in recent years throughout Europe, spread to new endemic foci, and became a notifiable disease in the European Union in 2012 [18,19]. The TBE vaccine licensed in Europe is recommended for all age groups in endemic areas with incidence rates exceeding 5 per 100,000 [19,20], however vaccination rates are very low [19]. Riccò et al. [18] conducted a knowledge, attitudes and practices survey of tick-borne encephalitis among occupational physicians in Italy, and found a lack of knowledge of TBE and low vaccine literacy among this stakeholder group. Improving knowledge of TBE, behavioral practices that prevent tick-bites, and vaccination would help prevent the spread of tick-borne infections such as TBE in Italy and elsewhere [18]. West Nile virus (WNV) has been circulating throughout Europe since the 1990s. A dramatic increase in WNV infections in multiple countries in Southern Europe was observed during 2018 in both humans and horses [21]. Castaldo et al. [22] provided a case report on two human patients from Italy. Clinical disease was characterized by atypical neurological presentation in these patients diagnosed with WNV neuroinvasive disease, involving the brainstem. The authors encouraged providers to remain alert to unusual disease presentations involving the central nervous system [22].

Author Contributions: Writing—original draft preparation, R.C.K.; writing—review and editing, A.C.B., J.D.B. All authors have read and agreed to the published version of the manuscript.

Funding: This research received no external funding.

Acknowledgments: Figure 1 was created using BioRender.com.

Conflicts of Interest: The authors declare no conflict of interest.

References

1. Jones, K.E.; Patel, N.G.; Levy, M.A.; Storeygard, A.; Balk, D.; Gittleman, J.L.; Daszak, P. Global trends in emerging infectious diseases. *Nature* **2008**, *451*, 990–993. [CrossRef] [PubMed]
2. Woolhouse, M.E.; Howey, R.; Gaunt, E.; Reilly, L.; Chase-Topping, M.; Savill, N. Temporal trends in the discovery of human viruses. *Proc. R. Soc. B Biol. Sci.* **2008**, *275*, 2111–2115. [CrossRef] [PubMed]
3. WHO Fact Sheet: Vector-Borne Diseases. Available online: https://www.who.int/news-room/fact-sheets/detail/vector-borne-diseases (accessed on 22 August 2020).
4. Rosenberg, R.; Johansson, M.A.; Powers, A.M.; Miller, B.R. Search strategy has influenced the discovery rate of human viruses. *Proc. Natl. Acad. Sci. USA* **2013**, *110*, 13961–13964. [CrossRef]
5. Weaver, S.C.; Reisen, W.K. Present and future arboviral threats. *Antivir. Res.* **2010**, *85*, 328. [CrossRef] [PubMed]
6. Young, K.I.; Medwid, J.T.; Azar, S.R.; Huff, R.M.; Drumm, H.; Coffey, L.L.; Pitts, R.J.; Buenemann, M.; Vasilakis, N.; Perera, D.; et al. Identification of mosquito bloodmeals collected in diverse habitats in malaysian borneo using COI barcoding. *Trop. Med. Infect. Dis.* **2020**, *5*, 51. [CrossRef] [PubMed]
7. Ab Hamid, N.; Mohd Noor, S.N.; Isa, N.R.; Md Rodzay, R.; Bachtiar Effendi, A.M.; Hafisool, A.A.; Azman, F.A.; Abdullah, S.F.; Kamarul Zaman, M.K.; Mohd Norsham, M.I.; et al. Vertical infestation profile of aedes in selected urban high-rise residences in Malaysia. *Trop. Med. Infect. Dis.* **2020**, *5*, 114. [CrossRef] [PubMed]
8. Julia da Silva Pessoa Vieira, C.; José Ferreira da Silva, D.; Rigotti Kubiszeski, J.; Ceschini Machado, L.; Pena, L.J.; Vieira de Morais Bronzoni, R.; da Luz Wallau, G. The emergence of Chikungunya ECSA lineage in a mayaro endemic region on the southern border of the Amazon Forest. *Trop. Med. Infect. Dis.* **2020**, *5*, 105. [CrossRef] [PubMed]
9. Magalhaes, T.; Chalegre, K.D.M.; Braga, C.; Foy, B.D. The endless challenges of arboviral diseases in Brazil. *Trop. Med. Infect. Dis.* **2020**, *5*, 75. [CrossRef] [PubMed]
10. Federal Select Agent Program. Available online: https://www.selectagents.gov/ (accessed on 16 January 2019).
11. Linthicum, K.J.; Britch, S.C.; Anyamba, A. Rift valley fever: An emerging mosquito-borne disease. *Annu. Rev. Entomol.* **2016**, *61*, 395–415. [CrossRef] [PubMed]
12. Grossi-Soyster, E.N.; LaBeaud, A.D. Rift valley fever: Important considerations for risk mitigation and future outbreaks. *Trop. Med. Infect. Dis.* **2020**, *5*, 89. [CrossRef] [PubMed]
13. Sorvillo, T.E.; Rodriguez, S.E.; Hudson, P.; Carey, M.; Rodriguez, L.L.; Spiropoulou, C.F.; Bird, B.H.; Spengler, J.R.; Bente, D.A. Towards a sustainable one health approach to crimean–congo hemorrhagic fever prevention: Focus areas and gaps in knowledge. *Trop. Med. Infect. Dis.* **2020**, *5*, 113. [CrossRef] [PubMed]
14. McLean, R.G. Letter to the editor: Venezuelan equine encephalitis virus 1B Invasion and epidemic control—South Texas, 1971. *Trop. Med. Infect. Dis.* **2020**, *5*, 104. [CrossRef] [PubMed]
15. Kading, R.C.; Cohnstaedt, L.W.; Fall, K.; Hamer, G.L. Emergence of arboviruses in the United States: The boom and bust of funding, innovation, and capacity. *Trop. Med. Infect. Dis.* **2020**, *5*, 96. [CrossRef] [PubMed]
16. Hinojosa, S.; Alquiza, A.; Guerrero, C.; Vanegas, D.; Tapangan, N.; Cano, N.; Olivarez, E. Detection of a locally-acquired zika virus outbreak in Hidalgo County, Texas through increased antenatal testing in a high-risk area. *Trop. Med. Infect. Dis.* **2020**, *5*, 128. [CrossRef] [PubMed]
17. Florida State Health Department Department of Health Daily Zika Update|Florida Department of Health. Available online: http://www.floridahealth.gov/newsroom/2017/01/012717-zika-update.html (accessed on 18 April 2020).
18. Riccò, M.; Gualerzi, G.; Ranzieri, S.; Ferraro, P.; Bragazzi, N.L. Knowledge, attitudes, practices (KAP) of Italian occupational physicians towards tick borne encephalitis. *Trop. Med. Infect. Dis.* **2020**, *5*, 117. [CrossRef] [PubMed]
19. Beauté, J.; Spiteri, G.; Warns-Petit, E.; Zeller, H. Tick-borne encephalitis in Europe, 2012 to 2016. *Eurosurveillance* **2018**, *23*. [CrossRef] [PubMed]
20. WHO. Vaccines against tick-borne encephalitis: WHO position paper. *Relev. Epidemiol. Hebd.* **2011**, *86*, 241–256.

21. Vilibic-Cavlek, T.; Savic, V.; Petrovic, T.; Toplak, I.; Barbic, L.; Petric, D.; Tabain, I.; Hrnjakovic-Cvjetkovic, I.; Bogdanic, M.; Klobucar, A.; et al. Emerging trends in the epidemiology of west nile and Usutu Virus infections in Southern Europe. *Front. Vet. Sci.* **2019**, *6*. [CrossRef] [PubMed]
22. Castaldo, N.; Graziano, E.; Peghin, M.; Gallo, T.; D'Agaro, P.; Sartor, A.; Bove, T.; Cocconi, R.; Merlino, G.; Bassetti, M. Neuroinvasive West Nile infection with an unusual clinical presentation: A single-center case series. *Trop. Med. Infect. Dis.* **2020**, *5*, 138. [CrossRef] [PubMed]

Editorial

The Endless Challenges of Arboviral Diseases in Brazil

Tereza Magalhaes [1,*], Karlos Diogo M. Chalegre [2], Cynthia Braga [3] and Brian D. Foy [1]

1 Arthropod-Borne and Infectious Diseases Laboratory (AIDL), Department of Microbiology, Immunology and Pathology (MIP), Colorado State University, Fort Collins, CO 80523-1692, USA; Brian.Foy@colostate.edu
2 Oswaldo Cruz Institute (IOC), Oswaldo Cruz Foundation (FIOCRUZ), Vice Presidency of Production and Innovation in Health (VPPIS), Rio de Janeiro, RJ 21040-900, Brazil; diogochalegre@gmail.com
3 Aggeu Magalhaes Institute (IAM), Oswaldo Cruz Foundation (FIOCRUZ), Department of Parasitology, Recife, PE 50670-420, Brazil; braga@cpqam.fiocruz.br
* Correspondence: Tereza.Magalhaes@colostate.edu

Received: 30 April 2020; Accepted: 7 May 2020; Published: 9 May 2020

Abstract: In this Editorial, we list and discuss some of the main challenges faced by the population and public health authorities in Brazil concerning arbovirus infections, including the occurrence of concurrent epidemics like the ongoing SARS-CoV-2/COVID-19 pandemic.

Keywords: dengue; zika; chikungunya; coronavirus; co-endemic

Optimal ecological and environmental conditions support year-long breeding of mosquito vectors of arboviruses in several Brazilian States. This, combined with socioeconomical factors that facilitate mosquito breeding (e.g., intermittent water supply that leads to short-term water storage in open-air artificial containers) and human exposure to mosquito bites, fosters cyclic and intense transmission of arboviruses in Brazil. In urban and peri-urban areas, the four dengue virus serotypes (DENV1-4), Zika virus (ZIKV), and chikungunya virus (CHIKV) are the most widespread and impactful mosquito-borne pathogens, all transmitted by the highly urbanized and anthropophagic *Aedes aegypti*. Arboviral diseases impose a great health burden to the population in Brazil and represent a constant challenge to health authorities. Diagnosis (and therefore clinical management) and notification of arboviral infections in co-endemic places (where more than one arbovirus co-circulate) are extremely complex. Point-of-care virus-specific testing is non-existent in the public and private health care sectors, and the most important diagnosis is clinical-epidemiological, upon which a case is notified to health authorities as suspected or confirmed based on the Ministry of Health case definitions (which uses clinical symptoms and blood test results, such as platelet counts). However, diseases caused by DENV, ZIKV and CHIKV can lead to similar acute symptoms that may differ only in time of onset, duration and severity [1]—thus, only well-trained, experienced physicians are more apt to correctly diagnose a patient, but even these professionals can misdiagnose without available virus-specific tests. Arboviral diseases are nationally notifiable diseases in Brazil, but in the public sector, which exclusively serves more than 70% of the Brazilian population, only a small proportion of cases notified by health care units undergo confirmatory tests in public reference laboratories through virus-specific molecular or serological assays, or virus culture. For instance, among the notified dengue cases in 2020 (until April), only approximately 23% were tested in reference laboratories [2]. In addition, for DENV and ZIKV, cross-reactivity of serological assays represents a serious issue as it can lead to erroneous results [3]. Official government notifications may thus be biased by inaccurate clinical diagnosis and cross-reactive serological results, and clinical management of infections may not be appropriate if the wrong diagnosis is made. The different socioeconomic realities of Brazilian States also contribute to inconsistencies of arboviral disease notifications.

The cross-reactivity between DENV and ZIKV serological assays are due to similar antigenic regions of viral proteins of these genetically related flaviviruses that can be recognized by the same antibodies. Besides being an issue in serological tests, cross-reactive DENV and ZIKV immunity can have important epidemiological implications in places where these viruses co-circulate. For instance, in vitro, in vivo and epidemiological studies have shown that pre-existing DENV immunity can either protect or enhance ZIKV infection, and consequently impact disease development [4–6]. Other studies suggest that the atypically low dengue incidence observed after the Zika epidemics in Brazil and other Latin American countries was due, in part, to short-term DENV protection from ZIKV infections [7,8]. Importantly, this lower dengue incidence was followed by a significant increase in dengue cases [2,7]. The impact of pre-existing DENV and ZIKV immunity in further heterologous infections and, importantly, in clinical diseases, needs to be continuously assessed in endemic areas.

It is also possible that ZIKV or other arboviruses may establish sylvatic transmission cycles in Brazil, as discussed by other authors [9]. If one looks at the map of Paulista, for example, a municipality within the Recife Metropolitan Region (RMR) in Pernambuco State that was heavily affected by ZIKV and CHIKV, forested areas surround all the urban areas where the viruses co-circulated and human cases were concentrated in 2015-16 (Figure 1 and [10]). These forested areas may harbor several sylvatic mosquitoes like *Aedes albopictus*, *Haemagogus janthinomys*, and *Sabethes tarsopus* that feed on non-human primates (NHPs) and may serve as vectors of arboviruses [11]. In addition, NHPs like the common marmoset *Callithrix jacchus* are abundant in the area [12] and found near humans. Importantly, ZIKV RNA and antibodies against several arboviruses have been found in NHPs in different regions of Brazil, including marmosets [13–16]. The seriousness of an established sylvatic arbovirus transmission cycle in NHPs and sylvatic mosquitoes in Brazil is well represented by yellow fever virus (YFV), which causes sporadic spillover human outbreaks leading to hundreds of deaths. Although a few studies have found little evidence of sylvatic ZIKV transmission in Brazil [16,17], the possibility of a sylvatic cycle being established in distinct Brazilian regions and at different times cannot be excluded. Further governmental or research-related arbovirus surveillance activities should intensify monitoring of sylvatic mosquitoes, NHPs and other small mammals, as the establishment of sylvatic cycles will require changes in the design of control programs.

The ZIKV outbreaks that occurred in Brazil in 2014-16 probably ceased due to herd immunity—however, instead of disappearing, the virus is still circulating in areas that were intensely affected, like the RMR, even if at low rates. In addition, virus transmission during the outbreaks was focal across metropolitan regions, where some areas were more intensely hit than others within the same municipality [4], corroborating the notion of clustered household/community transmission of arboviruses transmitted by *Ae. aegypti*. The low but constant circulation of ZIKV, the presence of prior virus foci with surrounding patchy areas containing higher numbers of naïve people, and the possibility of a sylvatic cycle being established in some regions increase the chances of unexpected re-emergence of the virus. It will also be important to assess the importance of sexual transmission among the sustained, low ZIKV circulation in endemic regions, as the epidemiological relevance of ZIKV sexual transmission may be higher than previously thought ([18] and Magalhaes et al., unpublished).

Escalating the problem of arboviral disease surveillance and management, concurrent outbreaks/epidemics of arboviruses and non-arthropod-borne pathogens can further complicate clinical diagnosis and completely overwhelm/saturate the health care system, as we may be seeing now with the pandemic of coronavirus disease (COVID-19) caused by severe acute respiratory syndrome coronavirus 2 (SARS-CoV-2). The number of notified dengue, Zika and chikungunya cases in Brazil in 2020 have reached over 660,000 by April [2], reflecting a difficult year for arboviral diseases in the country (Figure 2A). Although the true incidences of SARS-CoV-2 infections and COVID-19 cases are unknown in Brazil due to the very limited testing (currently, the Brazilian government recommends that only severe cases are tested in health clinics and hospitals), the notified numbers of infections and deaths are starting to increase, indicating a worsening epidemic scenario as of April 2020 (Figure 2B) [2]. At the moment, health care units like the local rapid-access units (Unidades de Pronto

Atendimento-UPAs), which serve communities like the Paulista population (~330,000 habitants), are working with a reduced number of staff as some individuals have fallen ill and many elderly professionals or those with comorbidities are on leave due to fear of becoming infected with the virus. In a recent serosurvey of SARS-CoV-2 antibodies among health professionals in Pernambuco State, 60% have tested positive, confirming these professionals are under very high risk of infection [19]. Although the highest numbers of notified arboviral diseases seemed to have occurred in March 2020, it is very likely that case notification has dropped as a result of fewer people infected with arboviruses seeking health facilities due to the SARS-CoV-2 pandemic. In fact, it would be important to see if household mosquito transmission of arboviruses increases because of social isolation during the COVID-19 pandemic, considering the endophilic behavior of *Ae. aegypti* (although social isolation is necessary, it is also important to assess its effects on other health factors). The blunt reality is that health care units have been dealing with a peak in arbovirus infections and COVID-19 cases concomitantly. Besides the many troubles inherent to an overwhelmed health care system, concurrent epidemics also can complicate clinical-epidemiological diagnoses. Some studies show that dengue cases can be misdiagnosed as respiratory infections and vice-versa [20,21]. Coinfections during concurrent epidemics must also be considered as they may worsen clinical diseases. Coinfections of influenza virus and DENV have been identified in several occasions during concurrent epidemics [22–24]. Future control efforts and programs must consider concurrent epidemics as they will most likely continue to happen in the future (e.g., epidemics of DENV and new strains of influenza virus).

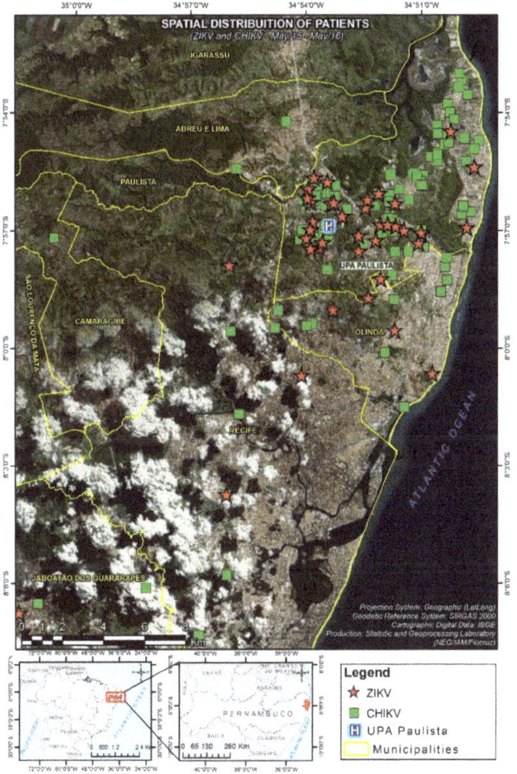

Figure 1. Mapped cases of Zika virus (ZIKV) and chikungunya virus (CHIKV) infections in Paulista, Pernambuco State, Brazil, in 2015-16. Note the green/forested areas surrounding the urban areas where cases were concentrated: an optimal interface for the establishment of sylvatic cycles of arbovirus transmission (this map was published in [10]).

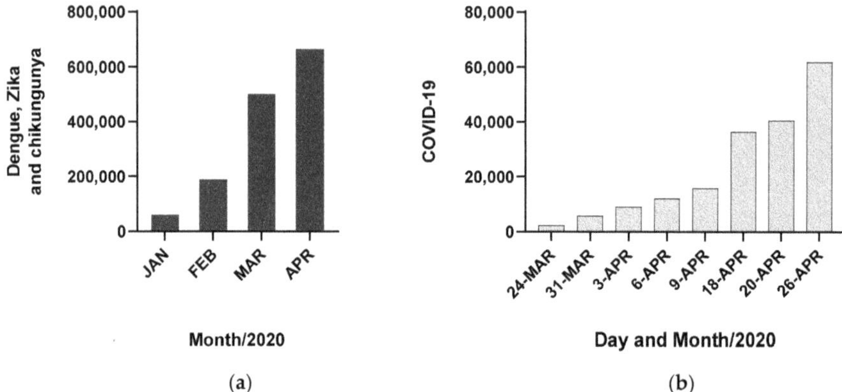

Figure 2. Cumulative notified cases of dengue, Zika and chikungunya (**a**) and COVID-19 (**b**) in Brazil as of April 2020.

Conclusions

Effective management of arboviral diseases in Brazil requires confronting major challenges. The co-endemicity of multiple and related arboviruses complicates clinical-epidemiological diagnoses, clinical management and case notification, in addition to impacting the epidemiology of arboviral diseases in unclear ways. The possible establishment of sylvatic transmission cycles will represent a significant additional challenge to the development of control programs and should be constantly monitored. Lastly, concurrent epidemics like the SARS-CoV-2/COVID-19 or other respiratory pathogens/illnesses can overwhelm health care systems and further complicate clinical-epidemiological diagnoses. Efforts to better control these diseases must seriously consider all these issues.

Author Contributions: Conceptualization, T.M.; investigation, T.M., K.D.M.C., C.B., B.D.F.; writing—original draft preparation, T.M., K.D.M.C., C.B., B.D.F.; writing—review and editing, T.M., K.D.M.C., C.B., B.D.F. All authors have read and agreed to the published version of the manuscript.

Funding: This research was funded by the National Institutes of Health (NIH, USA), grant number R21AI129464.

Conflicts of Interest: The authors declare no conflict of interest.

References

1. Brito, C.A.; Cordeiro, M.T. One year after the Zika virus outbreak in Brazil: From hypotheses to evidence. *Rev. Soc. Bras. Med. Trop.* **2016**, *49*, 537–543. [CrossRef] [PubMed]
2. Ministério da Saúde do Brasil. Boletins Epidemiológicos. Available online: https://www.saude.gov.br/boletins-epidemiologicos (accessed on 8 May 2020).
3. Zaidi, M.B.; Cedillo-Barron, L.; Gonzalez, Y.A.M.E.; Garcia-Cordero, J.; Campos, F.D.; Namorado-Tonix, K.; Perez, F. Serological tests reveal significant cross-reactive human antibody responses to Zika and Dengue viruses in the Mexican population. *Acta Trop.* **2020**, *201*, 105201. [CrossRef] [PubMed]
4. Rodriguez-Barraquer, I.; Costa, F.; Nascimento, E.J.M.; Nery, N.J.; Castanha, P.M.S.; Sacramento, G.A.; Cruz, J.; Carvalho, M.; De Olivera, D.; Hagan, J.E.; et al. Impact of preexisting dengue immunity on Zika virus emergence in a dengue endemic region. *Science* **2019**, *363*, 607–610. [CrossRef] [PubMed]
5. Oliveira, R.A.; de Oliveira-Filho, E.F.; Fernandes, A.I.; Brito, C.A.; Marques, E.T.; Tenorio, M.C.; Gil, L.H. Previous dengue or Zika virus exposure can drive to infection enhancement or neutralisation of other flaviviruses. *Mem. Inst. Oswaldo Cruz* **2019**, *114*, e190098. [CrossRef]
6. Watanabe, S.; Tan, N.W.W.; Chan, K.W.K.; Vasudevan, S.G. Dengue Virus and Zika Virus Serological Cross-reactivity and Their Impact on Pathogenesis in Mice. *J. Infect. Dis.* **2019**, *219*, 223–233. [CrossRef]

7. Borchering, R.K.; Huang, A.T.; Mier, Y.T.-R.L.; Rojas, D.P.; Rodriguez-Barraquer, I.; Katzelnick, L.C.; Martinez, S.D.; King, G.D.; Cinkovich, S.C.; Lessler, J.; et al. Impacts of Zika emergence in Latin America on endemic dengue transmission. *Nat. Commun.* **2019**, *10*, 5730. [CrossRef]
8. Perez, F.; Llau, A.; Gutierrez, G.; Bezerra, H.; Coelho, G.; Ault, S.; Barbiratto, S.B.; de Resende, M.C.; Cerezo, L.; Kleber, G.L.; et al. The decline of dengue in the Americas in 2017: Discussion of multiple hypotheses. *Trop. Med. Int. Health* **2019**, *24*, 442–453. [CrossRef]
9. Figueiredo, L.T.M. Human Urban Arboviruses Can Infect Wild Animals and Jump to Sylvatic Maintenance Cycles in South America. *Front. Cell. Infect. Microbiol.* **2019**, *9*, 259. [CrossRef]
10. Magalhaes, T.; Braga, C.; Cordeiro, M.T.; Oliveira, A.L.S.; Castanha, P.M.S.; Maciel, A.P.R.; Amancio, N.M.L.; Gouveia, P.N.; Peixoto-da-Silva, V.J., Jr.; Peixoto, T.F.L.; et al. Zika virus displacement by a chikungunya outbreak in Recife, Brazil. *PLoS Negl. Trop. Dis.* **2017**, *11*, e0006055. [CrossRef]
11. Aragao, N.C.; Muller, G.A.; Balbino, V.Q.; Costa Junior, C.R.; Figueiredo Junior, C.S.; Alencar, J.; Marcondes, C.B. A list of mosquito species of the Brazilian State of Pernambuco, including the first report of Haemagogus janthinomys (Diptera: Culicidae), yellow fever vector and 14 other species (Diptera: Culicidae). *Rev. Soc. Bras. Med. Trop.* **2010**, *43*, 458–459. [CrossRef]
12. Thompson, C.L.; Robl, N.J.; Melo, L.C.O.; Valença-Montenegro, M.M.; Valle, Y.B.M.; Oliveira, M.A.B.; Vinyard, C.J. Spatial distribution and exploitation of trees gouged by common marmosets (*Callithrix jacchus*). *Int. J. Primatol.* **2013**, *34*, 65–85. [CrossRef]
13. Terzian, A.C.B.; Zini, N.; Sacchetto, L.; Rocha, R.F.; Parra, M.C.P.; Del Sarto, J.L.; Dias, A.C.F.; Coutinho, F.; Rayra, J.; da Silva, R.A.; et al. Evidence of natural Zika virus infection in neotropical non-human primates in Brazil. *Sci. Rep.* **2018**, *8*, 16034. [CrossRef] [PubMed]
14. Favoretto, S.R.; Araujo, D.B.; Duarte, N.F.H.; Oliveira, D.B.L.; da Crus, N.G.; Mesquita, F.; Leal, F.; Machado, R.R.G.; Gaio, F.; Oliveira, W.F.; et al. Zika Virus in Peridomestic Neotropical Primates, Northeast Brazil. *Ecohealth* **2019**, *16*, 61–69. [CrossRef] [PubMed]
15. De Oliveira-Filho, E.F.; Oliveira, R.A.S.; Ferreira, D.R.A.; Laroque, P.O.; Pena, L.J.; Valenca-Montenegro, M.M.; Mota, R.A.; Gil, L. Seroprevalence of selected flaviviruses in free-living and captive capuchin monkeys in the state of Pernambuco, Brazil. *Transbound. Emerg. Dis.* **2018**, *65*, 1094–1097. [CrossRef] [PubMed]
16. Moreira-Soto, A.; Carneiro, I.O.; Fischer, C.; Feldmann, M.; Kummerer, B.M.; Silva, N.S.; Santos, U.G.; Souza, B.; Liborio, F.A.; Valenca-Montenegro, M.M.; et al. Limited Evidence for Infection of Urban and Peri-urban Nonhuman Primates with Zika and Chikungunya Viruses in Brazil. *mSphere* **2018**, *3*. [CrossRef]
17. Pauvolid-Correa, A.; Goncalves Dias, H.; Marina Siqueira Maia, L.; Porfirio, G.; Oliveira Morgado, T.; Sabino-Santos, G.; Helena Santa Rita, P.; Teixeira Gomes Barreto, W.; Carvalho de Macedo, G.; Marinho Torres, J.; et al. Zika Virus Surveillance at the Human-Animal Interface in West-Central Brazil, 2017–2018. *Viruses* **2019**, *11*, 1164. [CrossRef]
18. Rosenberg, E.S.; Doyle, K.; Munoz-Jordan, J.L.; Klein, L.; Adams, L.; Lozier, M.; Weiss, K.; Sharp, T.M.; Paz-Bailey, G. Prevalence and Incidence of Zika Virus Infection Among Household Contacts of Patients With Zika Virus Disease, Puerto Rico, 2016–2017. *J. Infect. Dis.* **2019**, *220*, 932–939. [CrossRef]
19. Centro de Informaçoes Estrategicas de Vigilância em Saúde de Pernambuco (CIEVS). Available online: https://www.cievspe.com/novo-coronavirus-2019-ncov (accessed on 8 May 2020).
20. Chacon, R.; Clara, A.W.; Jara, J.; Armero, J.; Lozano, C.; El Omeiri, N.; Widdowson, M.A.; Azziz-Baumgartner, E. Influenza Illness among Case-Patients Hospitalized for Suspected Dengue, El Salvador, 2012. *PLoS ONE* **2015**, *10*, e0140890. [CrossRef]
21. Restrepo, B.N.; Piedrahita, L.D.; Agudelo, I.Y.; Parra-Henao, G.; Osorio, J.E. Frequency and clinical features of dengue infection in a schoolchildren cohort from medellin, Colombia. *J. Trop. Med.* **2012**, *2012*, 120496. [CrossRef]
22. Perez, M.A.; Gordon, A.; Sanchez, F.; Narvaez, F.; Gutierrez, G.; Ortega, O.; Nunez, A.; Harris, E.; Balmaseda, A. Severe coinfections of dengue and pandemic influenza A H1N1 viruses. *Pediatr. Infect. Dis. J.* **2010**, *29*, 1052–1055. [CrossRef]

23. Lopez Rodriguez, E.; Tomashek, K.M.; Gregory, C.J.; Munoz, J.; Hunsperger, E.; Lorenzi, O.D.; Irizarry, J.G.; Garcia-Gubern, C. Co-infection with dengue virus and pandemic (H1N1) 2009 virus. *Emerg. Infect. Dis.* **2010**, *16*, 882–884. [CrossRef] [PubMed]
24. Hussain, R.; Al-Omar, I.; Memish, Z.A. The diagnostic challenge of pandemic H1N1 2009 virus in a dengue-endemic region: A case report of combined infection in Jeddah, Kingdom of Saudi Arabia. *J. Infect. Public Health* **2012**, *5*, 199–202. [CrossRef] [PubMed]

Tropical Medicine and Infectious Disease

Article

Detection of a Locally-Acquired Zika Virus Outbreak in Hidalgo County, Texas through Increased Antenatal Testing in a High-Risk Area

Steven Hinojosa *, Alexander Alquiza, Clarissa Guerrero, Diana Vanegas, Niko Tapangan, Narda Cano and Eduardo Olivarez

Hidalgo County Health and Human Services Department, Edinburg, TX 78542, USA; alexalquiza1@gmail.com (A.A.); Clarissa.Guerrero@hchd.org (C.G.); Diana.Vanegas@hchd.org (D.V.); Niko.Tapangan@hchd.org (N.T.); Narda.Cano@hchd.org (N.C.); Eddie.Olivarez@hchd.org (E.O.)

* Correspondence: Steven.Hinojosa@hchd.org; Tel.: +1-956-318-2426 (ext. 7344)

Received: 29 June 2020; Accepted: 3 August 2020; Published: 5 August 2020

Abstract: Hidalgo County (HC), located along the Texas–Mexico border, was listed as a high-risk county for Zika virus (ZIKV) in 2017 by the Texas Department of State Health Services, based on its historical presence of Dengue. Due to its subtropical climate, active binational travel, and population of low socioeconomic status, Hidalgo County focused on disease detection activities for the prevention of further transmission. Therefore, Hidalgo County Health and Human Services enacted public health surveillance, reviewed laboratory results, and conducted epidemiological investigations from 2016 to 2018. In 2017, Hidalgo County experienced a locally-acquired outbreak of Zika virus disease, resulting in the highest local mosquito-borne acquisition case count for the year within the United States. This resulted in Hidalgo County reviewing epidemiological data for disease detection and risk areas. With the data review, key outcomes of testing were identified. This included the importance of both RT-PCR and IgM-ELISA/PRNT testing methods. In addition, increased antenatal testing and surveillance also recognized the need of improved disease identification and testing among the general population, especially during localized outbreaks.

Keywords: Zika; tropical disease; epidemiology; border health; outbreak

---

1. Introduction

Hidalgo County fosters a transient, unique community supplemented by common travel along the Texas–Mexico border and multiple demographic factors that culminate in a high-risk population for arboviral diseases [1]. The majority of the Hidalgo County population is Hispanic or Latin at 91.8% (770,794) [2,3]. The median age is 28.9 and it has the 5th largest adolescent population in Texas. A pronounced proportion of the population was in poverty at 31.8%, more than double the national average (14.6%) and nearly double the average of Texas (16.0%). Furthermore, the median household income was well below both the state and national median by about USD 20,000 per year. Also, a great part (31.6%) of the civilian noninstitutionalized population of Hidalgo County lived with no health insurance, triple the rate of the country and nearly double the rate of the state. In addition, there was an educational disparity in Hidalgo County with only 63.7% of the population (25 and older) possessing a high school degree or higher; the state's and country's numbers are well above 80%. Over half of the population was female (51.1%) with about 42.2% of them between ages 15 and 44 [2,3]. The birth rate in Hidalgo County of women in this age range is 5.78% whereas the state and national averages are 5.65% and 5.21%, respectively. Lastly, Hidalgo County fosters a tropical environment suitable for the presence of the mosquito species *Aedes aegypti* and *Aedes albopictus*, the primary vectors of the Zika virus [4,5].

In April 2017, the Texas Department of State Health Services (DSHS) released a health alert, urging increased testing of pregnant women and symptomatic individuals in high-risk counties, resulting in a large spike in Zika virus testing. Recommendations were set forth by the Texas DSHS for residents of counties within the Lower Rio Grande Valley (i.e., Cameron, Hidalgo, Starr, Webb, Willacy, and Zapata) [6]. This included testing of pregnant women in their first and second trimester as a part of their prenatal care, and of symptomatic individuals presenting with a rash and at least one other common Zika virus (ZIKV) symptom, either fever, joint pain, or conjunctivitis (red eye) [5]. In April 2018, this health alert was updated to include additional high-risk counties in addition to the previous counties (i.e., Kinney, Maverick, and Val Verde counties). In this 2018 update, asymptomatic pregnant women were recommended they be tested three times during their pregnancy, once per trimester using RT-PCR only instead of paired RT-PCR and IgM-ELISA samples as recommended in 2017. The general population testing recommendation stayed the same in regards to symptom-based testing, encompassing rash and one other symptom associated with ZIKV (i.e., joint pain, conjunctivitis, fever) [7].

Hidalgo County is comprised of a younger population (over 5 years below both the state and national medians) with education and income disparities in addition to frequent cross-border travel and inherent environmental risks. Hidalgo County has also had historical evidence of Dengue transmission, another arboviral illness transmitted by the mosquito species *Aedes aegypti* and *Aedes albopictus* [8,9]. Thus, the Hidalgo County population presents as a vulnerable population with concerns of high-risk Zika virus transmission. Due to these risk factors within this population, local testing efforts for ZIKV increased and an outbreak of local acquisition was identified in the later part of 2017. This resulted in Hidalgo County having the highest local mosquito-borne acquisition case count, with four of the total seven symptomatic disease cases within the United States for 2017 [10]. This surveillance report aimed at pinpointing Zika virus cases, understanding routes of transmission, and analyzing the representation of different sources and types of testing. With this information, the goal was to establish and quantify the most effective testing measures to prevent spread of the virus. From there, implications on future surveillance and policies would better serve to protect the community in a preventative capacity to stop any locally-acquired outbreaks from occurring. High-yield tests and testing sources would be emphasized to create maximum outreach. Additionally, the data obtained and conclusions formed may assist in further scientific research conducted on similar communities.

2. Materials and Methods

A surveillance report was conducted to assess ZIKV testing outcomes of patients in Hidalgo County, Texas. The work conducted was considered non-research Public Health Surveillance; therefore, it was not subject to IRB review requirements. Electronic lab results were submitted to the National Electronic Disease Surveillance System (NEDSS, v5.4.6-GA, Centers for Disease Control and Prevention, Atlanta, GA, USA) by laboratories performing Zika virus tests on serum and urine samples. Commercial laboratories automatically sent positive and presumptive positive IgM-ELISA specimens to Centers of Disease Control and Prevention (CDC) laboratories for confirmation by plaque-reducing neutralization testing (PRNT). This PRNT method would assess antibodies of Dengue-1, Dengue-2, and ZIKV to assess possible cross-reactivity. CDC laboratories would then submit these laboratory results to the NEDSS database. Hidalgo County Zika virus staff ran and exported a report to abstract data from 1 January 2016 to 31 December 2018, reflecting all electronic ZIKV lab reports for this time period.

All cases were defined as either positive or negative. Then, each was designated a test type as IgM-ELISA, RT-PCR, or PRNT. Furthermore, the data were categorized into specimen type as serum, urine, or other (including umbilical blood and seminal fluid) as well as the facility type as hospital, obstetrics and gynecology (OB-GYN), unknown OB-GYN, private clinic, and public health.

Microsoft Excel (v2016, Microsoft Corp., Redmond, WA, USA) and StataIC (v15.1, 64-bit, StataCorp LLC, College Station, TX, USA) programs were used to analyze the extracted data from NEDSS. These data were evaluated with descriptive statistics on the total number of cases, median age,

age range, percentage of female patients, percentage of patients tested by OB/GYNs, travel history, and the change in tests ordered in 2017 and 2018 after a Texas DSHS Zika virus Health Advisory updated testing recommendations for pregnant women to once per trimester. Texas DSHS travel history was defined as no travel history within the last 12 months since date of collection. Descriptive statistics were also used to identify the different case characteristics from the 2017 Zika virus outbreak.

3. Results

Testing for ZIKV among patients in Hidalgo County began in June 2016. During this time period, until 31 December 2018, a total number of 29,045 lab results came from a total of 15,016 patients. Some patients received multiple lab tests in compliance with DSHS recommendations to test during each trimester. The median age of these patients was 27 years old ($n = 29{,}014$), excluding 31 cases where date of birth was not collected. The age range of these lab results ranged from birth (0 days) to 99 years old. The majority of specimens (55.5%) were RT-PCR testing, as noted in Table 1. The vast majority (98.31%) of patients tested were female patients ($n = 28{,}569$), shown in Table 2. With new testing recommendations and capacity, testing increased 54.3 times more in 2017 compared to 2016.

Table 1. Zika virus testing by test type, 2016–2018. This table depicts testing by test type and total positives by test type. Plaque-reducing neutralization testing (PRNT) took place for IgM non-negative specimens. IgM samples had a 2.18% positivity rate, while RT-PCR samples had a 0.11% positivity rate. Overall positivity rate for these specimens was 1.16%.

	IgM-ELISA	RT-PCR	PRNT	Total Results
2016	70 (25.36%)	196 (71.01%)	10 (3.62%)	276
2017	6649 (45.50%)	7855 (53.75%)	110 (0.75%)	14,614
2018	6015 (42.49%)	8026 (56.70%)	114 (0.81%)	14,155
Total Tested	12,734 (43.84%)	16,077 (55.35%)	234 (0.81%)	29,045
Total Positives	277 (2.18%)	17 (0.11%)	44 (18.80%)	338 (1.16%)

Table 2. Gender representation of Zika virus specimen testing by facility type from 2016 to 2018. Testing with OB-GYN clinics was identified as the most common facility type for testing both females and males. Males had the lowest testing count at private laboratories (without a facility) and private clinics. Males represented only 1.69% of the specimens tested.

	Source of Testing (2016–2018)					
	OB-GYN	Hospital	Lab, no Facility	Private Clinic	Public Health	Total
Female	23,807	1545	43	1694	1465	28,554 (98.31%)
Male	225	100	8	58	100	491 (1.69%)
Total	24,032 (82.74%)	1645 (5.67%)	51 (0.18%)	1752 (6.03%)	1565 (5.39%)	29,045

In 2017, a total of 14,614 specimens were processed for residents in Hidalgo County. Figure 1 shows the 2017 trends in testing outcomes by month, where an increase in ZIKV testing and non-negative test results were witnessed, beginning April 2017, after the Texas DSHS Health Advisory. Of the specimens tested, 110 of the 117 presumptive positive laboratory specimens were forwarded to the CDC for confirmation. As per Texas DSHS epidemiological case criteria, CDC confirmation was needed to check for false positives and cross-reactivity. Of these forwarded presumptive results, 22 samples were confirmed as positive by PRNT, while all other of these IgM sample results were identified as false-positive. This resulted in 16 unique and unduplicated ZIKV patients. An additional two RT-PCR cases were also positive, but they were not subject to CDC confirmation due to their confirmatory status. These cases were divided into two categories, as defined by Texas DSHS, Zika disease cases and Zika infection cases. Zika disease cases were defined as having clinical symptomatic evidence of the virus along with laboratory confirmation. Zika virus infection cases were defined as having laboratory confirmation without clinical symptomatic evidence. In Table 2, the breakdown of case type (symptom presence), acquisition, and pregnancy status is described. Of the 18 cases, eight were symptomatic disease and ten were asymptomatic infection; eleven of these cases were identified as travel-associated (61.11%). Furthermore, most of the cases involved a pregnant patient (72.22%). Only five of the confirmed cases were identified as locally-acquired, where no international travel history was identified within the last 12 months, which indicates potential transmission within the Hidalgo community. Only two cases' origins remained undetermined, as shown in Table 3. The five locally-acquired cases occurred in four different cities, which included McAllen, Pharr, Alamo, and Mercedes, Texas. These communities did not present similar geographical or demographical findings, presenting an outbreak within multiple local communities, rather than one specific neighborhood.

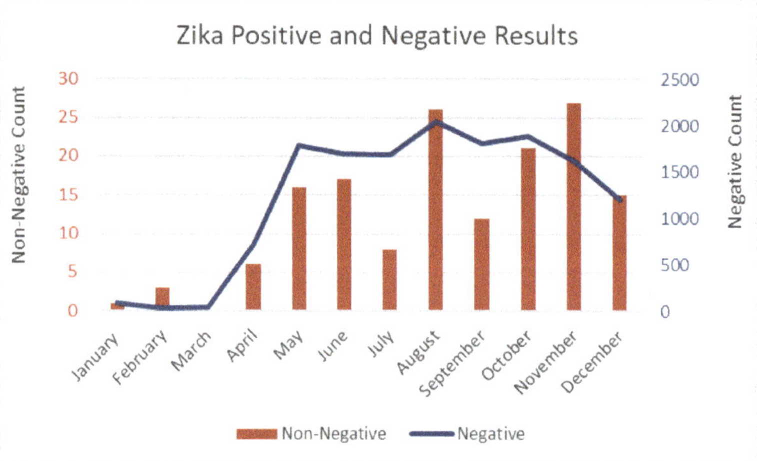

Figure 1. The line depicts a sharp increase in the number of Zika virus (ZIKV) specimens tested (noted as a blue line) and non-negative results (noted as red bars), after the 2017 Zika virus Health Alert was released. A total of 14,614 labs were processed, resulting in 150 non-negative samples. Non-negative samples included results that were positive, equivocal, indeterminate, or other results that were not negative. Testing peaked in August 2017, with 26 non-negative samples and 2036 negative samples.

Table 3. This table presents the breakdown of cases that were symptomatic disease cases versus asymptomatic infections. Acquisition was also determined, with 5 cases identified as locally- acquired. The majority of cases were pregnant with 75% pregnant symptomatic disease cases and 70% asymptomatic infection cases.

2017 Zika Case Count	Travel-Associated	Locally-Acquired	Acquisition Undetermined	Total	Pregnant
Zika Disease Cases	4	4	0	8	6 out of 8
Zika Infection Cases	7	1	2	10	7 out of 10

Of these 18 cases, ages ranged from 6 to 56 years, with an average age of 24.2 years and a median of 22.0 years. When assessing gender, 2 cases were male and 16 were female. In regard to travel, seven cases had travel history to neighboring Mexico cities in Tamaulipas, three cases to Nuevo Leon, and one case to both southern Mexico and Honduras. Of these 18 cases, 16 were identified as serum IgM-ELISA presumptive positive, along with CDC PRNT positive results, and 2 samples were positive through urine and serum via RT-PCR.

In 2018, testing stayed consistent with 14,155 specimens. RT-PCR testing slightly increased, while IgM-ELISA testing slightly decreased, as shown in Table 1. This included a total of nine Zika virus cases. One case was symptomatic disease and eight were asymptomatic infections. Three cases were identified as local acquisition, one case with unknown acquisition, and five cases identified as travel-associated. All cases were pregnant at the time of testing. However, in 2019, testing began to show a decrease with a total of 9323 results, with three Zika virus cases identified. All three cases were asymptomatic infection and were pregnant. One case had unknown acquisition and the other two cases were identified as travel-associated.

4. Discussion

Increased efforts of testing, through the Texas Zika virus public health alert and provider education, assisted in the identification of local acquisition ZIKV cases within Hidalgo County, Texas. Prior to this health advisory, asymptomatic testing for Zika virus was not routine. Therefore, with this advisory and new testing recommendations, asymptomatic Zika infections, which may not have originally been identified, were detected. In addition, with 75% of Zika disease cases and 70% of Zika asymptomatic infection cases being pregnant at the time of testing, this poses a question as to what could be the true case representation of ZIKV cases in the non-pregnant population. Therefore, increasing both surveillance efforts and disease awareness are necessary for the medical community, especially at times of outbreaks, in regards to testing and identification of cases within the general population. In addition, the majority of the cases (16 of the 18) were serology-based testing, accompanied by PRNT confirmation. Therefore, exposure time periods may be less clear. However, with multiple infections occurring in the later part of 2017, exposure may have taken place around this time. Serology-based testing produced a higher rate of false-positives after CDC testing confirmed these samples to be actually negative. However, although RT-PCR testing is the recommended method, without these serology-based testing methods, these cases may have never been identified through routine provider testing and public health surveillance. Therefore, utilization of serology, paired with PRNT testing to account for Dengue cross-reactivity, may be beneficial in future continued surveillance efforts, especially during times of increased arboviral activity within the community, compared to PCR alone. This testing method is helpful during times of Dengue outbreaks, where outbreaks of local acquisition were identified in Hidalgo County in 2013 and 2019. Furthermore, OB/GYN providers testing partners of pregnant females provided a strong surveillance system for identification of infection, and assisted in preventative education for negative partners. In addition, new testing recommendations also assisted in the detection of an asymptotic male partner of a pregnant female. Although this outbreak was limited in the number of cases with local acquisition, these findings present the importance of continued surveillance efforts along the southern Texas–Mexico border for Zika virus and other mosquito-borne illnesses.

Author Contributions: Conceptualization, S.H., C.G. and D.V.; Data curation, C.G., N.T. and N.C.; Formal analysis, S.H. and A.A.; Funding acquisition, E.O.; Investigation, D.V., N.T. and N.C.; Methodology, C.G.; Project administration, S.H. and E.O.; Supervision, S.H., D.V. and E.O.; Writing—original draft, S.H., A.A., C.G., D.V., N.T. and N.C.; Writing—review and editing, S.H., A.A., C.G. and D.V. All authors have read and agreed to the published version of the manuscript.

Funding: This effort was funded in part by the Texas DSHS Zika Epidemiology and Laboratory Capacity (ELC) grant, Texas DSHS Local Public Health Services (LPHS) grant, and award 1U01CK000512 01 from the Centers for Disease Control and Prevention to establish the Western Gulf Center of Excellence for Vector-Borne Diseases.

Acknowledgments: The authors thank all participants of this Zika surveillance project. This includes, but is not limited to the epidemiology staff at Hidalgo County Health and Human Services Department and the Zoonosis Control Branch staff Texas Department of State Health Services. The authors would also like to thank the Centers for Disease Control and Prevention (CDC) and their staff for conducting the laboratory confirmation testing of samples.

Conflicts of Interest: The authors declare no conflict of interest. In addition, the funders had no role in the design of the surveillance report; in the collection, analyses, or interpretation of data; in the writing of the manuscript, or in the decision to publish the results.

References

1. Shacham, E.; Nelson, E.J.; Hoft, D.F.; Schootman, M.; Garza, A. Potential High-Risk Areas for Zika Virus Transmission in the Contiguous United States. *Am. J. Public Health* **2017**, *107*, 724–731. [CrossRef] [PubMed]
2. American FactFinder. Available online: https://www.census.gov/quickfacts/fact/table/hidalgocountytexas,US/HEA775217#HEA775217 (accessed on 30 April 2020).
3. American FactFinder. Available online: https://factfinder.census.gov/faces/tableservices/jsf/pages/productview.xhtml?pid=ACS_17_1YR_S1301&prodType=table (accessed on 30 April 2020).
4. Champion, S.R.; Vitek, C.J. Aedes aegypti and Aedes albopictus Habitat Preferences in South Texas, USA. *Environ. Health Insights* **2014**, *8*, 35–42. [CrossRef] [PubMed]
5. McKenzie, B.A.; Wilson, A.E.; Zohdy, S. Aedes albopictus is a competent vector of Zika virus: A meta-analysis. *PLoS ONE* **2019**, *14*, e0216794. [CrossRef] [PubMed]
6. Texas Department of State Health Services. (n.d.). Zika Health Alert—7 April 2017. Available online: https://www.dshs.state.tx.us/news/releases/2017/HealthAlert-04072017.aspx (accessed on 22 June 2020).
7. Texas Department of State Health Services. (n.d.). Zika Health Alert—2 April 2018. Available online: https://www.dshs.texas.gov/news/releases/2018/HealthAlert-04022018.aspx (accessed on 22 June 2020).
8. Bhatt, S.; Gething, P.W.; Brady, O.J. The global distribution and burden of dengue. *Nature* **2013**, *496*, 504–507. [CrossRef] [PubMed]
9. Hotez, P.J. The rise of neglected tropical diseases in the "new Texas". *PLOS Neglect. Trop. Dis.* **2018**, *12*, e0005581. [CrossRef] [PubMed]
10. Centers for Disease Control and Prevention. 2017 Case Counts in the US. Available online: https://www.cdc.gov/zika/reporting/2017-case-counts.html (accessed on 24 April 2019).

Article

Knowledge, Attitudes, Practices (KAP) of Italian Occupational Physicians towards Tick Borne Encephalitis

Matteo Riccò [1,*], Giovanni Gualerzi [2], Silvia Ranzieri [3], Pietro Ferraro [4] and Nicola Luigi Bragazzi [5]

1 AUSL–IRCCS di Reggio Emilia, Occupational Health and Safety Service on the Workplaces of the Local Health Unit of Reggio Emilia (SPSAL), Via Amendola n.2, I-42122 Reggio Emilia, Italy
2 Department of Medicine and Surgery, School of Medicine, University of Parma, Via Gramsci n.14, 43123 Parma, Italy; gualerzi@gmail.com
3 Department of Medicine and Surgery, School of Occupational Medicine, University of Parma, Via Gramsci n.14, I-43123 Parma, Italy; silvia.ranzieri@studenti.unipr.it
4 ASL Foggia, Department of Prevention, Occupational Health and Safety Service on the Workplaces of the Local Health Unit of Foggia, Piazza Pavoncelli 11, I-41121 Foggia, Italy; dott.pietro.ferraro@gmail.com
5 Laboratory for Industrial and Applied Mathematics (LIAM), Department of Mathematics and Statistics, York University, 4700 Keele Street, Toronto, ON M3J 1P3, Canada; bragazzi@yorku.ca
* Correspondence: mricco2000@gmail.com or matteo.ricco@ausl.re.it

Received: 25 May 2020; Accepted: 14 July 2020; Published: 16 July 2020

Abstract: Tick-Borne Encephalitis (TBE) is an occupational health threat with increasing incidence in the geographic area of Italy. Despite this, TBE vaccination rates have ranged from 10% to 40% in Italy, even in at-risk workers. The reasons for this low rate are investigated in this present study of the knowledge, attitudes, and practices of occupational physicians (OP) regarding TBE disease and vaccination in at-risk workers. A total of 229 OP participated in an internet-based survey by completing a structured questionnaire. Adequate general knowledge of TBE disease was found in 58% of OP. Accurate perception of TBE risk in occupational settings was found in 20%. TBE vaccination for at-risk workers was recommended by 19%. Willingness to recommend TBE vaccination was more likely by OP practicing in endemic areas (Odds Ratio 3.10, 95% confidence intervals 1.47–6.55), who knew the existence of the term "arboviruses" (3.10, 1.29–7.44), or exhibited a better understanding of TBE (2.38, 1.11–5.12)—and were positive predictors for promoting TBE vaccine, while acknowledging that TBE as a severe disease was a negative one. Tick-borne disorders in Italy are a still rare (but increasing) occupational health threat, and vaccination gaps for TBE virus may find an explanation in OP incomplete knowledge of evidence-based recommendations.

Keywords: *Ixodes ricinus*; knowledge; risk perception; tick; tick-borne encephalitis (TBE); occupational physicians

1. Introduction

Tick-borne encephalitis (TBE) is a potentially lethal vaccine-preventable infection of the central nervous system (CNS) caused by an arbovirus (TBE virus, or TBEV) included within the Flavivirus genus, Flaviviridae family [1]. Overall incidence rates for TBE in most Western European countries have ranged between 0.4 and 0.6/100,000 over the years 2011–2015; but notification rates in certain areas (e.g., Czech Republic, Estonia, Lithuania, Slovenia) have largely exceeded 5 cases/100,000 population/years [2–5].

With mortality rates up to 2% and long-term neurological complications in around 10% of patients, TBE has emerged as an increasing cause of morbidity and mortality [1–3,6–9]. The geographical distribution of the competent vector, *Ixodes* spp., is expanding as an effect of climate change. For this reason, previously spared countries have become endemic for TBE [8,10,11], particularly in northern regions and in mountainous territory, whose forests support the life cycle of *Ixodes ricinus* [1–3,8,9].

For instance, Italy has historically been considered as a relatively low-risk region, having no cases reported until 1978, and a total of 456 cases notified between 2000 and 2016 [2,3,7]. With 0.1 incident cases/100,000 population/year and no TBE-related deaths since 2016, Italy remains, actually, a low-incidence country [4]. Since 2015, a total of 135 cases have been reported; the majority of them clustered in North-Eastern Subalpine Regions of Trentino-South Tyrol, Friuli-Venezia-Giulia, and Veneto [3,4,12]. Recent estimates suggest that some smaller areas in the aforementioned regions of Trentino (1.0 case/100,000 inhabitants for the province of Trento as a whole during 2000–2013, peaking to 41.6 in the municipality of Tres), Friuli-Venezia-Giulia (1.0 case/100,000 inhabitants for the whole province of Udine, but 181.2/100,000 for the community of Tramonti di Sopra in the time period 2000–2013) [3,12], and Veneto (e.g., 5.95/100,000 population in the province of Belluno 2007–2018, peaking to 35.9/100,000 in the municipality of Limana in 2000–2013), [2] largely exceed national TBE estimates [3,8,9,12,13], and could be classified as highly endemic areas (i.e., >5 cases/100,000 inhabitants/year) according to current World Health Organization (WHO) definition [1–3,6–9,12].

As a consequence, TBE has become a significant health threat for outdoor workers in North Eastern Italy [6,7,14–16], and the Italian Ministry of Health has recommended TBE vaccination for forestry workers, farmers, and veterinarians from endemic areas [1–3,8,12]. Even though official data on TBE vaccination rates in Italy are lacking, some reports hint towards figures ranging between 10% and 40%, even in high-risk groups from highly endemic areas [2–4,7,12]. Such recommendations have been issued by means of the National Immunization Prevention Plan 2017–2019 (Piano Nazionale Prevenzione Vaccinale—PNPV), a guidance document for immunization policies [17,18]. Application of immunization policies for workplaces is a main issue for occupational physicians, the medical professionals responsible for health promotion in workplaces [18–21]. More precisely, Italian occupational physicians are not only requested to properly tailor national recommendations to specific occupational settings, but they should recall the vaccination status of the assisted workers, informing them about the pros and cons of recommended vaccinations [18,20,21]. Moreover, occupational physicians (OP) should actively inform workers about occupational risk factors, including vectors and microorganisms. As prevention of TBE may also benefit from several environmental (e.g., pesticide treatment of the working environment) and behavioral (e.g., promoting self-assessment for tick bites at the end of outdoor activities) interventions, the assessment of knowledge (i.e., the awareness of official recommendations), attitudes (i.e., propensity towards a certain intervention), and practices (i.e., actual application of such intervention)—or knowledge, attitudes, practices (KAP)—of occupational physicians towards TBE and TBE vaccine, they can be useful in order to improve the health and safety of high-risk outdoor workers [10,14,22].

Although studies about outdoor workers have assessed their KAP toward TBE preventive measures [10,14,23–26], occupational physicians have been scarcely investigated. Therefore, the main endpoint of this study is to assess KAP of occupational physicians concerning TBE preventive measures, particularly TBE vaccination, in order to assess factors associated with TBE vaccine acceptancy, and alternative or complementary preventive practices.

2. Materials and Methods

Study design. A cross-sectional questionnaire-based study was performed between 01/02/2020 and 28/02/2020, involving occupational physicians participating in seven different private Facebook group pages and four closed forums, focusing on occupational medicine, and whose applications were officially limited to occupational physicians. The group pages had approximately 2034 unique members, but no information could be obtained regarding cross-inscriptions, not even how many of these members were actively using Facebook.

To post the study invitation, the chief researcher contacted the administrators, requesting preventive authorization to post the link to the questionnaire, including a short description of the aims of the survey. Facebook users who clicked on the invitation texts were provided with the full study information, an opportunity to give their informed consent, and a web link to the survey (Google Forms; Google LLC; Menlo Park, CA, USA). The survey was conducted in Italian. To be included in the sample, the participant was supposed to be living and working in Italy as an occupational physician. If a potential participant was found not to match the inclusion criteria, the survey closed down. The survey was anonymous, and no personal data, such as name, IP address, email address, or personal information unnecessary to the survey, was requested, saved, or tracked. No monetary or other compensation was offered to the participants.

Questionnaire. The questionnaire was formulated in Italian, and its test–retest reliability was preventively assessed through a survey on 10 occupational physicians completing the questionnaire at two different points in time. The testing questionnaires were ultimately excluded from the final analyses. All questions were self-reported, and not externally validated. An English translation of the questionnaire is available on request from the corresponding author. The final questionnaire included the following sections:

1. individual characteristics: age (by decennial groups), sex, whether they had encountered at least a TBE case in their practice (yes vs. no), and the Italian Region where the professional mainly worked and lived. The latter factor was eventually dichotomized as endemic vs. non-endemic for TBE;
2. knowledge test: participants were initially requested whether they knew the meaning of the term "arbovirus", being able to explain it. Participants then received a knowledge test containing a set of 20 true–false statements, elaborated through extensive literature review, covering typical misconceptions about arboviruses (e.g., "TBE vaccine is effective also against Lyme disease"; FALSE) [5,9,23–30]. Across the knowledge test, TBE was always reported as "TBE" or "tick-borne encephalitis" (in English), avoiding the Italian translation (i.e., "Encefalite da zecca"). A General Knowledge Score (GKS) was then calculated as the sum of correctly and incorrectly marked recommendations: when the participants answered correctly, +1 was added to a sum score, whereas a wrong indication or a missing/"don't know" answer added 0 to the sum score. GKS was then dichotomized by median value in higher vs. lower knowledge status;
3. risk perception: participants were initially asked to rate the perceived severity (C^{INF}) and the perceived frequency (I^{INF}) of TBE in agricultural and forestry settings by means of a fully labeled 5-points Likert scale. The available options ranged from "not significant" (i.e., "of no significant concern in daily practice", score 1) to "very significant" (i.e., "of very high concern in daily practice", score 5). As perceived risk has been defined as a function of the perceived probability of an event and its expected consequences [21,27], a Risk Perception Score (RPS) was eventually calculated as follows:

$$RPS = I^{INF} \times C^{INF} \quad (1)$$

4. attitudes and practices: we inquired participants whether they recommend TBE vaccine for high-risk groups (yes vs. no). A series of possible interventions for prevention of arboviral infections in workplaces were then reported to the participants, and they were asked to report which ones they perceived as useful in order to prevent TBE, the tick-borne Lyme disease, and the mosquito-borne arboviral infection West Nile Fever (WNF). Namely (useful vs. useless): removal of standing water

from the working environment; treatment of standing water with chemicals and/or biological agents; use of pesticides; use of light-colored cloths; use of full-length trousers; putting the end of trousers into the socks; self-assessment at the end of the outdoor activities.

Ethical considerations. Before giving their consent to the survey, participants were briefed that all information would be gathered anonymously and handled confidentially. Participation was voluntary, and the questionnaire was collected only from subjects who had expressed consent for study participation. As individual participants cannot be identified based on the presented material, this study caused no plausible harm or stigma to participating individuals. As the study had an anonymous, observational design, and did not include clinical data about patients, nor configured itself as a clinical trial, a preliminary evaluation by an Ethical Committee was not required, according to the Italian law (Gazzetta Ufficiale no. 76, dated 31/3/2008).

Data analysis. Continuous variables were initially tested for normal distribution (D'Agostino and Pearson omnibus normality test): where the corresponding p value was < 0.10, "normal" distribution was assumed as rejected, and variables were compared through Mann–Whitney or Kruskal–Wallis tests for multiple independent samples. On the other hand, variables passing the normality check (D'Agostino and Pearson p value ≥ 0.10) were compared using the Student's t test or ANOVA, where appropriate. Categorical variables were reported as per cent values, and their distribution in respect of the outcome variable of promoting TBE vaccine was initially analyzed through chi-squared test. The paired proportions of preventive measures for TBE vs. Lyme disease, and TBE vs. the WNF were also tested through the McNemar test, assuming as the null hypothesis that no differences were reported.

All categorical variables that at univariate analysis were significantly associated with a positive attitude towards TBE vaccine (i.e., $p < 0.05$) were included in a stepwise binary logistic regression analysis model in order to calculate adjusted odds ratios (aOR) and their respective 95% confidence intervals (95%CI). All statistical analyses were performed by means of IBM SPSS Statistics 24.0 for Macintosh (IBM Corp. Armonk, NY, USA).

3. Results

3.1. Descriptive Analysis

As shown in Table 1, a total of 229 occupational physician (11.3% of the eligible population) participated to the inquiry. The majority of respondents were aged 40 years or more (66.9%); 52.0% were males, and 48.0% females. Overall, 41.9% of participants came from Northern regions (i.e., Valle d'Aosta, Piemonte, Liguria, Lombardia, Veneto, Trentino-Südtirol, Friuli-Venezia-Giulia, Emilia-Romagna), 40.6% came from Central Italy (i.e., Toscana, Umbria, Marche, Abruzzo, Lazio), and residual 17.5% from Southern regions (i.e., Campania, Molise, Puglia, Basilicata, Calabria) and major islands of Sicilia and Sardinia. Of all respondents, 44.5% resided in Italian regions characterized by incident cases of TBE, and 10.9% had any previous interaction with at least one patient affected by TBE.

3.2. Assessment of Knowledge about TBE

After normalization, the mean GKS was generally low (58.4% ± 12.0; actual range 30.0%–80.0%; median 60.0%). Internal consistency coefficient amounted to Cronbach's alpha = 0.707 (Table 2).

Table 1. Characteristics of 229 Italian occupational physicians participating in the survey (2020). Likert scale for perceived severity and perceived frequency of tick-borne encephalitis (TBE) in agricultural and forestry settings were dichotomized as "significant" and "very significant" (i.e., severe and frequently reported disease) vs. all other values (i.e., not severe and infrequently reported).

	Variable	No., %	Average ± SD
Age Group			
	<30 years	10, 4.4%	
	30–39 years	66, 28.8%	
	40–49 years	102, 44.5%	
	50–59 years	38, 16.6%	
	≥60 years	13, 5.7%	
Gender			
	Male	119, 52.0%	
	Female	110, 48.0%	
Residence			
	Northern Italy	96, 41.9%	
	Central Italy	93, 40.6%	
	Southern Italy	40, 17.5%	
	Residence in Italian Region Endemic for TBE	102, 44.5%	
	Knowledge of the Term "arbovirus"	143, 62.4%	
	Any Previous Interaction with TBE case(s) in the Practice	25, 10.9%	
	TBE Immunization Recommended in High-Risk Occupational Groups	44, 19.2%	
	General Knowledge Score		58.4% ± 12.0 (median 60.0%)
	General Knowledge Score > Median (i.e., 60.0%)	79, 34.5%	
	TBE Acknowledged as a Severe Disease	38, 16.6%	
	TBE Acknowledged as a Frequently Reported Disease	77, 33.6%	
	Risk Perception Score		20.7% ± 12.8 (median 16.0%)

Notes: TBE = Tick Borne Encephalitis; Northern Italy = Italian regions of Valle d'Aosta, Piemonte, Liguria, Lombardia, Veneto, Trentino-Südtirol, Friuli-Venezia-Giulia, Emilia-Romagna; Central Italy = Italian regions of Toscana, Umbria, Marche, Abruzzo, Lazio; Southern Italy = Italian regions of Campania, Molise, Puglia, Basilicata, Calabria, Sicilia, Sardinia.

Interestingly, only 16.6% of respondents were aware that the Italian region where they live/work is characterized by incident TBE cases, while around a quarter of participants (27.9%) acknowledged TBE as a vaccine preventable disease, and only 27.1% understood that TBE vaccine is not active against Lyme disease. Moreover, participants were affected by uncertainties about transmission of arboviruses, including the very same TBE (i.e., only 55.5% reporting tick bite as instrumental to human infection), but also dengue (only 7.0% clearly stated that dengue virus cannot be transmitted by tick bite) and Crimean-Congo hemorrhagic fever (14.4% correctly reporting tick as the infection's primary vector). Significant uncertainties involved also the immediate managing of tick bite, as 25.3% reported that oils/lotions may improve the removal of tick head.

Table 2. Knowledge test: response distribution of presented items proposed to the 229 Occupational Physicians participating in the survey and contributing to the assessment of General Knowledge Score (GKS) (Cronbach's Alpha = 0.707).

Statement	Correct Answer	No., %
1. The subsequent disorders are transmitted by tick bite		
TBE	TRUE	127, 55.5%
Crimea-Congo Hemorrhagic Fever	TRUE	33, 14.4%
Lyme Disease	TRUE	173, 75.5%
Yellow Fever	FALSE	219, 96.5%
Dengue	FALSE	16, 7.0%
West Nile Fever	FALSE	223, 97.4%
2. The subsequent disorders are preventable through commercially available vaccination(s)		
TBE	TRUE	64, 27.9%
Crimea-Congo Hemorrhagic Fever	FALSE	198, 86.5%
Lyme Disease	FALSE	205, 89.5%
Yellow Fever	TRUE	130, 56.8%
Dengue	FALSE	175, 76.4%
West Nile Fever	FALSE	183, 79.9%
3. In the Italian Region where he/she lives/works TBE is endemic	*	38, 16.6%
4. In cases of tick bites, the head should be removed as soon as possible	TRUE	184, 80.3%
5. Tick head should be removed by means of specifically designed tweezers	TRUE	157, 68.6%
6. Tick head removal may be improved by means of oils/lotions	FALSE	58, 25.3%
7. TBE vaccine is effective also against Lyme disease	FALSE	62, 27.1%
8. Arboviral infections in at-risk professionals are compensated as occupational injuries	TRUE	123, 53.7%
9. TBE is characterized by inter-human spreading	FALSE	174, 76.0%
10. Latency for Lyme's disease may be of weeks up to some months	TRUE	135, 59.0%

Note = * depending of the region of residence.

3.3. Assessment of Attitudes and Practices

A total of 44 respondents acknowledged that they usually recommend TBE vaccination in agricultural and forestry workers, i.e., 19.2% of total sample, with only 68.8% of respondents aware that the disorder may be prevented by means of vaccination.

Focusing on the reported preventive measures (Figure 1), TBE was associated with a distinctive pattern, even when compared to a similarly tick-borne disorder such as Lyme disease. Even though very few respondents addressed preventive treatment of standing waters, more properly reported for preventing a mosquito-borne infection such as WNF (7.9% vs. 64.2% for the removal of standing water, and 3.9% vs. 67.7% for chemical/biological treatment; both comparisons $p < 0.001$), such interventions were associated with an effective prevention of Lyme disease (16.2% for removal of standing water, and 13.5% for any other treatment, $p = 0.001$ and $p < 0.001$ compared to TBE). A significant share of respondents largely overlooked otherwise effective interventions against tick bites, such as the use of repellents (41.0% for TBE and 47.6% for Lyme disease), and the use of appropriate pesticides (35.8% and 22.3%, for TBE and Lyme disease, respectively). Both repellents and pesticides were more commonly reported against WNF (69.4%, and 40.2%); even though only for the use of repellents, the difference was statistically significant. The highly effective behavioral adaptation of promoting self-assessment at the end of outdoor activities was reported by 69.9% of respondents as effective against TBE, by 63.3% of them against Lyme disease ($p = 0.044$), and by 40.2% against WNF ($p < 0.001$). Still, we identified a mixed understanding of other behavioral adaptations, as a huge proportion of respondents advocated for TBE only countermeasures effective also against Lyme disease, such as wearing full-length trousers (87.3% vs. 66.4% for Lyme disease), and putting the end of trousers into the socks (70.7% vs. 54.6%). Less than one fifth of participants apparently promoted the use of light-colored cloths for TBE prevention (19.2%), a share significantly lower than those reported for Lyme disease (61.1%, $p < 0.001$) and even WNF (34.5%, $p < 0.001$).

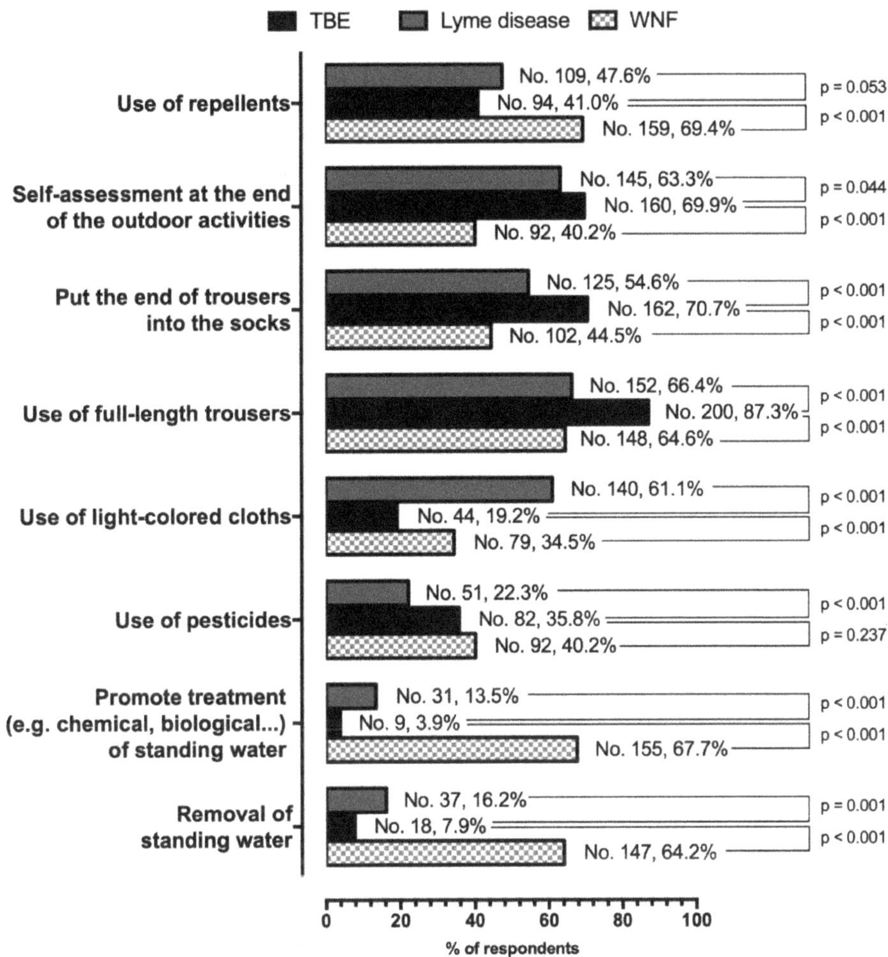

Figure 1. Prevalence of protective practices for TBE, Lyme disease, and West Nile Fever (WNF) reportedly recommended by 229 Occupational Physicians participating in the survey, compared by means of McNemar test for paired proportions.

3.4. Assessment of the Risk Perception

Overall, 16.6% of respondents acknowledged TBE severity as significant/highly significant, while 33.6% of them similarly reported its frequency as significant/highly significant. The large majority of respondents, then, did not characterize TBE as a disease of significant severity and occurrence in daily practice.

A correspondent RPS equals to 20.7% ± 12.8 (range 4.0 to 60.0%) was calculated, stressing a diffuse underestimation of TBE on the workplaces.

3.5. Univariate Analysis

Interestingly, RPS and GKS were well correlated ($r = 0.304$, $p < 0.001$): a better knowledge status (i.e., fewer misconceptions and/or less personal attitudes guiding the vaccine decisions) was associated with a greater risk perception for TBE infection. Univariate analysis showed that promotion of TBE vaccine among agricultural and forestry workers was significantly associated with being from areas characterized by incident TBE cases (65.9% vs. 39.5% in professionals from non-endemic areas, $p = 0.003$), knowing and understanding the term "arbovirus" (81.8% vs. 57.8%, $p = 0.005$), scoring a higher GKS (50.0% in subjects with GKS > median vs. 30.8% for GKS ≤ median, $p = 0.026$). On the contrary, a proactive attitude was less frequently reported among occupational physicians acknowledging TBE as a severe disease than among those underscoring its severity (2.3% vs. 20.0%, $p = 0.009$).

3.6. Multivariate Analysis

In regression analysis (Table 3), being from areas endemic for TBE (aOR 3.107, 95%CI 1.473 to 6.553), knowing the term "arbovirus" (aOR 3.104, 95%CI 1.295 to 7.442), and reporting a better GKS (aOR 2.386, 95%CI 1.112 to 5.120) were identified as positive predictors for a proactive attitude towards TBE vaccination. On the contrary, acknowledging TBE as a severe disease was identified a negative effector (aOR 0.068; 95%CI 0.009–0.524).

Table 3. Analysis of factors associated with promoting tick-borne encephalitis (TBE) vaccine in 229 Italian Occupational Physicians (2020).

Variable	TBE Immunization Recommended in High-Risk Workers		Chi Squared Test *p* Value	aOR (95% CI)
	Yes (No./44, %)	No (No./185, %)		
Age < 40 Years	13, 29.5%	63, 34.1%	0.695	-
Male Gender	18, 40.9%	101, 54.6%	0.143	-
Residence in Northern Italy vs. Other Regions	24, 54.5%	72, 38.0%	0.086	-
Residence in Italian Region Endemic for TBE	29, 65.9%	73, 39.5%	0.003	3.107 (1.473; 6.553)
Knowledge of the Term "arbovirus"	36, 81.8%	107, 57.8%	0.005	3.104 (1.295; 7.442)
Any Previous Interaction with TBE case(s) in the Practice	5, 11.4%	20, 10.8%	1.000	-
General Knowledge Score > Median	22, 50.0%	57, 30.8%	0.026	2.386 (1.112; 5.120)
TBE Acknowledged as a Severe Disease	1, 2.3%	37, 20.0%	0.009	0.068 (0.009; 0.524)
TBE Acknowledged as a Frequently Reported Disease	11, 25.0%	66, 35.7%	0.242	-

Notes: aOR = adjusted Odds Ratio (i.e., Odds Ratio calculated through binary logistic regression).

4. Discussion

In our survey, the large majority of respondents was apparently unaware that a relatively uncommon but severe disease such as TBE may be avoided through an effective and reliable vaccine. As a consequence, only a small share of sampled Italian occupational physicians exhibited willingness to recommend the TBE vaccine, a feature that was substantially consistent with the relatively low acceptance for TBE vaccine even among high-risk occupational groups [20]. Main factors involved in TBE vaccine promotion among high-risk groups (i.e., agricultural and forestry workers) were identified in residing in high-risk areas, having a certain understanding about arboviruses, and exhibiting a better knowledge of TBE and tick-borne diseases. Not coincidentally, study participants exhibited a diffuse misunderstanding of TBE and its preventive measures, with risk perception significantly associated with the assessed knowledge status.

Such results were not unexpected. On the one hand, the Health Belief Model points out that personal experiences and knowledge status are the logical prerequisites to raise risk perception, which in turn promotes behavioral adaptations [22,28]. On the other hand, Occupational Physicians acknowledging TBE as a severe disease were less frequently promoting TBE vaccine among high-risk groups. These results may appear as inconsistent, or even openly in opposition with the basic assumption that a better understanding of the risk associated with a certain disorder will ultimately increase the acceptance of countermeasures [22,28], but it should be stressed that with an overall annual incidence of 0.1 cases/100,000 people, TBE remains an uncommon event for the majority of professionals not working in high-risk areas [1–5]. Previous studies on KAP of Italian OP have shown significant knowledge gaps and misunderstandings on up-to-date vaccinology [18,20]. Moreover, international reports hint towards diffuse false beliefs on immunization practice among such healthcare professionals [21]. Interestingly, while some previous reports have inquired KAP of medical professionals on other tick-borne disorders (e.g., Lyme disease) [29–31], there is a significant lack of evidence on OP and their understanding of TBE [7,14–16,32]. Some studies that have examined acceptance of preventive measures in endemic areas found uneven or low use of risk-reducing measures, suggesting that people at risk either ignore or underestimate the health threat from TBE, being unaware of risks and protective measures, or do not understand that the adoption of protective measures is worth their time [14,15,33–36].

Still, some aspects of the diffuse misunderstandings on TBE and its prevention were particularly frustrating, for the following reasons. First, it is obvious that while TBE may be easily prevented by avoiding high-risk habitats during the peak period of tick activity [1,9,31,37], agricultural and forestry workers may be forced to perform their tasks irrespective of such place and time restrictions. Second, TBE and Lyme disease are two distinctive disorders, for their etiology, and pathological aspects, but their prevention may benefit from the very same preventive habits. Unfortunately, sampled occupational physicians seemly assumed TBE and Lyme disease as more distinctive entities, with the potential consequence of compromising the prevention of other tick-borne viral infections, such as Crimea-Congo hemorrhagic fever, but also bacterial (e.g., *Rickettsia* spp., *Anaplasma phagocytophilum*), and protozoan agents (e.g., *Babesia microti*). Third, it should be stressed that occupational physicians are quite effective in implementing acceptance and knowledge among other high-risk workers [18,20]. Addressing knowledge gaps of occupational physicians may then maximize the consent for vaccination programs and the overcome of mutual misunderstanding between public health professionals and vaccine hesitant individuals or even vaccine objectors.

The promotion of appropriate preventive measures, either behavioral or immunization, among Italian agricultural and forestry workers, is indispensable, as some hints highlight that knowledge gaps and reported misbeliefs on TBE/TBE vaccine issues may have actively contributed to the inappropriately low protective rates otherwise recalled in previous international reports [16,24,25,32,38].

First, not only overall acceptance [20], but also national vaccination rates for TBE among high-risk workers remain quite low [2,7,9,12,14], even though serological studies have pointed out that infections do occur, also in Italian regions not usually associated with TBE [7,12,13,15,16]. Interestingly, in some previous studies, the most commonly reported reason for not being vaccinated against TBE was the unawareness that a vaccine exists [20], despite a quite good interaction with healthcare providers such as the general practitioner and the occupational physician [14,31]. In other words, while our results seemingly confirm the earlier suspect that healthcare professionals (particularly occupational physicians) are not actively promoting TBE immunization in high–risk workers groups, since these professionals show a diffuse neglect of such prevention strategy, vaccination rates may be significantly improved by enriching their knowledge about TBEV and TBE vaccine [7,14,20,31].

Second, a previous report from North-Eastern Italy pointed out that a high percentage of farmers usually avoid behavioral protective and preventive measures [14,33,35]. Despite half of participants recalled tick bites while performing their work tasks [14], the systematic body checks for tick bites after field work (i.e., 2.8%) was often not performed [33,36,38,39]. Keeping in mind the increasing threat represented by TBE in some Italian Regions, but also that *Borrelia burgdorferi* actively circulates among Italian isolates of *Ixodes ricinus*, and reliable vaccines for Lyme disease still remain unavailable, an appropriate intervention of occupational physicians on such specific topic has, therefore, the potential to significantly improve health and safety of agricultural and forestry workers. Despite the fact that TBEV infection occurs in the first minutes after a tick bite, the virus begins to multiply actively in the feeding tick, increasing the infection dose, and, in turn, increasing the risk of disease. This means that the rapid removal of ticks reduces the risk not only for bacterial infections, but also for TBE [40]. For this reason, understanding and sharing appropriate methods for the removal of attached ticks may extensively reduce the overall risk for the majority of tick borne infections [34,41].

Our study shows several limitations. Firstly, it shares the implicit limits of Internet-based surveys [42,43]. Such studies are actually reliable, cost-effective, and mostly much faster than a paper-based survey. However, as participants are somehow "self-selected", the final sample may potentially over-represent some sub-groups (e.g., subjects from younger age groups, with a greater literacy, and more accustomed to the internet access), eventually failing to represent the original population. Therefore, a significant selection bias cannot be ruled out. Again, as in conventional paper-based surveys, participating voluntarily could be due to a proactive attitude or greater knowledge about vaccinations. In the same way, the fact of not participating could be understood as a negative attitude or a lack of knowledge about vaccinations.

Again, we cannot rule out that some of the items assessed through the knowledge test may be affected by a significant social desirability bias, with participants reporting the "socially appropriated" rather than their authentic behaviors, so that our results could have ultimately overstated the share of occupational physicians having an effective understanding of TBE associated issues [14,44,45].

Moreover, our sample was of limited size, including only 229 professionals among the over 7000 occupational physicians from the national list [20]. Even though we required the participants to report their geographic origin, such factor was assessed among the possible effectors for TBE-related KAP, without any role in the sampling strategy. As Italy has been repetitively acknowledged for very heterogeneous health literacy and vaccination rates, our results should be cautiously interpreted as representative of the national level [17,18,46]. Similarly, while a certain selection is usually performed by social media managers of specific discussion groups (e.g., by registering only subjects who receive a specific invitation by the manager; answering to specific "selection" questions; etc.), we cannot rule out that some of the respondents were not actively working as occupational physicians, limitedly or even not fulfilling our initial selection criteria.

Finally, collected data were not externally validated, lacking an estimate of high-risk workers actually followed by sampled occupational physicians. More specifically, we are neither able to ascertain how often sampled professionals interact with agricultural and/or forestry workers, nor which share of their practice they actually represent. Similarly, we are unable to assess how reliable are the practices reported by respondents, that is which share of workers followed by participants actually receive vaccines and/or specific recommendations. As a consequence, we were unable to estimate the effective extent of the social desirability bias, being the actual vaccination rates for TBE potentially even lower than those self-reported by study participants.

5. Conclusions

In conclusion, our results suggest an extensive lack of knowledge of sampled occupational physicians on TBE vaccination. Such results are consistent with previous reports on the TBE and its preventive measures in Northern Italy, and with the limited evidence on vaccine literacy of occupational physicians. More specifically participants were seemly unaware that an effective TBE vaccine does exist, and underestimated the potential health risks associated with TBEV infection. Our results suggest that a significant share of occupational physicians actually ignores or only partially applies official recommendations on TBE vaccinology. As knowledge status appeared well correlated with risk perception, being moreover characterized as a significant predictor of a proactive attitude towards TBE among occupational physicians, it is reasonable that filling their information gaps might improve the rate of proper preventive measures among agricultural and forestry workers. As TBEV infection may be effectively countered not only by means of TBE vaccine, but also through effective behavioral practices, improving this way the prevention of all tick-borne infections, increasing vaccination literacy of occupational physicians and promoting their interaction with agricultural and forestry workers would be, therefore, instrumental and cost-effective in reducing the potential spreading of all tick-borne disorders.

Author Contributions: Conceptualization, M.R. and N.L.B.; methodology, M.R.; software, M.R., G.G.; validation, G.G.; formal analysis, M.R.; investigation, G.G., P.F., S.R.; resources, P.F., M.R.; data curation, M.R.; writing—original draft preparation, M.R., G.G., S.R.; writing—review and editing, S.R.; visualization, M.R., F.P., N.L.B., S.R.; supervision, M.R., N.L.B.; project administration, M.R.; funding acquisition, N/A. All authors have read and agreed to the published version of the manuscript.

Funding: This research received no external funding.

Conflicts of Interest: The authors declare no conflict of interest. The facts, conclusions, and opinions stated in the article represent the authors' research, conclusions, and opinions, and are believed to be substantiated, accurate, valid, and reliable. However, as this article includes the results of personal researches of the authors, presenting correspondent, personal conclusions, and opinions, parent employers are not forced in any way to endorse or share its content and its potential implications.

References

1. Riccardi, N.; Antonello, R.M.; Luzzati, R.; Zajkowska, J.; Di Bella, S.; Giacobbe, D.R. Tick-borne encephalitis in Europe: A brief update on epidemiology, diagnosis, prevention, and treatment. *Eur. J. Intern. Med.* **2019**, *62*, 1–6. [CrossRef]
2. Cocchio, S.; Bertoncello, C.; Napoletano, G.; Claus, M.; Furlan, P.; Fonzo, M.; Gagliani, A.; Saia, M.; Russo, F.; Baldovin, T.; et al. Do we know the true burden of tick-borne encephalitis? A cross-sectional study. *Neuroepidemiology* **2020**, *54*, 227–234. [CrossRef] [PubMed]
3. Rezza, G.; Farchi, F.; Pezzotti, P.; Ruscio, M.; Lo Presti, A.; Ciccozzi, M.; Mondardini, V.; Paternoster, C.; Bassetti, M.; Merelli, M.; et al. Tick-borne encephalitis in North-East Italy: A 14-year retrospective study, January 2000 to December 2013. *Eurosurveillance* **2015**, *20*, 30034. [CrossRef] [PubMed]
4. European Centre for Disease Prevention and Control, Tick-Borne Encephalitis Annual Epidemiological Report for 2018 Key Facts. 2019. Available online: https://www.ecdc.europa.eu/sites/default/files/documents/TBE-annual-epidemiological-report-2018.pdf (accessed on 18 December 2019).
5. Beauté, J.; Spiteri, G.; Warns-Petit, E.; Zeller, H. Tick-borne encephalitis in europe, 2012 to 2016. *Eurosurveillance* **2018**, *23*, 1800201. [CrossRef] [PubMed]
6. Cinco, M.; Barbone, F.; Ciufolini, M.G.; Mascioli, M.; Rosenfeld, M.A.; Stefanel, P.; Luzzati, R. Seroprevalence of tick-borne infections in forestry rangers from northeastern Italy. *Clin. Microbiol. Infect.* **2004**, *10*, 1056–1061. [CrossRef]
7. di Renzi, S.; Martini, A.; Binazzi, A.; Marinaccio, A.; Vonesch, N.; D'Amico, W.; Moro, T.; Fiorentini, C.; Ciufolini, M.G.; Visca, P.; et al. Risk of acquiring tick-borne infections in forestry workers from Lazio, Italy. *Eur. J. Clin. Microbiol. Infect. Dis.* **2010**, *29*, 1579–1581. [CrossRef]

8. Baráková, I.; Derdáková, M.; Selyemová, D.; Chvostáč, M.; Špitalská, E.; Rosso, F.; Collini, M.; Rosà, R.; Tagliapietra, V.; Girardi, M.; et al. Tick-borne pathogens and their reservoir hosts in northern Italy. *Ticks Tick. Borne. Dis.* **2017**, *9*, 164–170. [CrossRef] [PubMed]
9. Amicizia, D.; Domnich, A.; Panatto, D.; Lai, P.L.; Cristina, M.L.; Avio, U.; Gasparini, R. Epidemiology of tick-borne encephalitis (TBE) in Europe and its prevention by available vaccines. *Hum. Vaccines Immunother.* **2013**, *9*, 1163–1171. [CrossRef]
10. Arikan, I.; Kasifoglu, N.; Metintas, S.; Kalyoncu, C. Knowledge, beliefs, and practices regarding tick bites in the Turkish population in a rural area of the Middle Anatolian Region. *Trop. Anim. Health Prod.* **2010**, *42*, 669–675. [CrossRef]
11. Alkishe, A.A.; Peterson, A.T.; Samy, A.M. Climate change influences on the potential geographic distribution of the disease vector tick Ixodes ricinus. *PLoS ONE* **2017**, *12*, e0189092. [CrossRef]
12. Dipartimento di Prevenzione-Azienda Provinciale Per i Servizi Sanitari (APSS) Della Provincia Autonoma di Trento. MALATTIE TRASMESSE DA ZECCHE IN TRENTINO Aggiornamento a Giugno 2019 [in Italian: Tick Born Diseases in Trentino Province-Update June 2019]. Available online: https://www.apss.tn.it/documents/10180//443975//Malattie+trasmesse+da+zecche+-+aggiornamento+giugno+2019 (accessed on 22 May 2020).
13. Pistone, D.; Pajoro, M.; Novakova, E.; Vicari, N.; Gaiardelli, C.; Viganò, R.; Luzzago, C.; Montagna, M.; Lanfranchi, P. Ticks and bacterial tick-borne pathogens in Piemonte region, Northwest Italy. *Exp. Appl. Acarol.* **2017**, *73*, 477–491. [CrossRef]
14. Riccò, M.; Bragazzi, N.L.; Vezzosi, L.; Balzarini, F.; Colucci, E.; Veronesi, L.; Riccò, M.; Bragazzi, N.L.; Vezzosi, L.; Balzarini, F.; et al. Knowledge, attitudes, and practices on tick-borne human diseases and tick-borne encephalitis vaccine among farmers from North-Eastern Italy (2017). *J. Agromed.* **2020**, *25*, 73–85. [CrossRef] [PubMed]
15. Tomao, P.; Ciceroni, L.; D'Ovidio, M.C.; de Rosa, M.; Vonesch, N.; Iavicoli, S.; Signorini, S.; Ciarrocchi, S.; Ciufolini, M.G.; Fiorentini, C.; et al. Prevalence and incidence of antibodies to Borrelia burgdorferi and to tick-borne encephalitis virus in agricultural and forestry workers from Tuscany, Italy. *Eur. J. Clin. Microbiol. Infect. Dis.* **2005**, *24*, 457–463. [CrossRef] [PubMed]
16. Vonesch, N.; Binazzi, A.; Bonafede, M.; Melis, P.; Ruggieri, A.; Iavicoli, S.; Tomao, P. Emerging zoonotic viral infections of occupational health importance. *Pathog. Dis.* **2019**, *77*, ftz018. [CrossRef] [PubMed]
17. Bonanni, P.; Ferrero, A.; Guerra, R.; Iannazzo, S.; Odone, A.; Pompa, M.; Rizzuto, E.; Signorelli, C. Vaccine coverage in Italy and assessment of the 2012-2014 National Immunization Prevention Plan. *Epidemiol. Prev.* **2015**, *39*, 146–158.
18. Durando, P.; Dini, G.; Massa, E.; La Torre, G. Tackling biological risk in the workplace: Updates and prospects regarding vaccinations for subjects at risk of occupational exposure in Italy. *Vaccines* **2019**, *7*, 141. [CrossRef]
19. Manzoli, L.; Sotgiu, G.; Magnavita, N.; Durando, P.; Barchitta, M.; Carducci, A.; Conversano, M.; de Pasquale, G.; Dini, G.; Firenze, A.; et al. Evidence-based approach for continuous improvement of occupational health. *Epidemiol. Prev.* **2015**, *39*, 81–85.
20. Riccò, M.; Cattani, S.; Casagranda, F.; Gualerzi, G.; Signorelli, C. Knowledge, attitudes, beliefs and practices of occupational physicians towards vaccinations of health care workers: A cross sectional pilot study in north-eastern Italy. *Int. J. Occup. Med. Environ. Health* **2017**, *30*, 775–790. [CrossRef]
21. Betsch, C.; Wicker, S. Personal attitudes and misconceptions, not official recommendations guide occupational physicians' vaccination decisions. *Vaccine* **2014**, *32*, 4478–4484. [CrossRef]
22. Fall, E.; Izaute, M.; Baggioni, N.C. How can the health belief model and self-determination theory predict both influenza vaccination and vaccination intention? A longitudinal study among university students. *Psychol. Health* **2017**, *33*, 746–764. [CrossRef] [PubMed]
23. Beaujean, D.J.M.A.; Gassner, F.; Wong, A.; Crutzen, R.; Ruwaard, D. Determinants and protective behaviours regarding tick bites among school children in the Netherlands: A cross-sectional study. *BMC Public Health* **2013**, *13*, 1148. [CrossRef]

24. Butler, A.D.; Sedghi, T.; Petrini, J.R.; Ahmadi, R. Tick-borne disease preventive practices and perceptions in an endemic area. *Ticks Tick-Borne Dis.* **2016**, *7*, 331–337. [CrossRef] [PubMed]
25. Cych, P.; Kiewra, D.; Szczepańska, A. Prevention and the state of knowledge of tick- borne diseases among orienteers in Poland prevention and the state of knowledge of tick-borne diseases among orienteers in Poland. *Infect. Dis.* **2018**, *50*, 236–238. [CrossRef]
26. Zöldi, V.; Turunen, T.; Lyytikäinen, O.; Sane, J. Knowledge, attitudes, and practices regarding ticks and tick-borne diseases, Finland. *Ticks Tick-Borne Dis.* **2017**, *8*, 872–877. [CrossRef] [PubMed]
27. Yates, F.J.; Stone, E.R. The Risk Construct. In *Risk-Taking Behaviour*; Yates, F.J., Ed.; John Wiley & Sons: Chichester, UK, 1992; pp. 1–25. ISBN 0471922501.
28. Piacentino, J.D. Occupational risk of Lyme disease: An epidemiological review. *Occup. Environ. Med.* **2002**, *59*, 75–84. [CrossRef]
29. Henry, B.; Crabtree, A.; Roth, D.; Blackman, D.; Morshed, M. Lyme disease: Knowledge, beliefs, and practices of physicians in a low-endemic area. *Can. Fam. Physician* **2012**, *58*, 289–295.
30. Ferrouillet, C.; Milord, F.; Lambert, L.; Vibien, A.; Ravel, A. Lyme disease: Knowledge and practices of family practitioners in southern Quebec. *Can. J. Infect. Dis. Med. Microbiol.* **2015**, *26*, 151–156. [CrossRef] [PubMed]
31. Vandererven, C.; Bellanger, A.P.; Faucher, J.F.; Marguet, P. Prise en charge des morsures de tiques par les médecins généralistes en Franche-Comté, 2013. *Med. Mal. Infect.* **2017**, *47*, 261–265. [CrossRef]
32. Vonesch, N.; D'Ovidio, M.C.; Melis, P.; Remoli, M.E.; Ciufolini, M.G.; Tomao, P. Climate change, vector-borne diseases and working population. *Ann Ist Super Sanità* **2016**, *52*, 397–405. [CrossRef]
33. Slunge, D.; Boman, A. Learning to live with ticks? The role of exposure and risk perceptions in protective behaviour against tick-borne diseases. *PLoS ONE* **2018**, *13*, e0198286. [CrossRef]
34. Slunge, D. The willingness to pay for vaccination against tick-borne encephalitis and implications for public health policy: Evidence from Sweden. *PLoS ONE* **2015**, *10*, e0143875. [CrossRef] [PubMed]
35. Jones, T.F.; Garman, R.L.; LaFleur, B.; Stephan, S.J.; Schaffner, W. Risk factors for tick exposure and suboptimal adherence to preventive recommendations. *Am. J. Prev. Med.* **2002**, *23*, 47–50. [CrossRef]
36. Stjernberg, L.; Berglund, J. Tick prevention in a population living in a highly endemic area. *Scand. J. Public Health* **2005**, *33*, 432–438. [CrossRef] [PubMed]
37. Piesman, J.; Eisen, L. Prevention of tick-borne diseases. *Annu. Rev. Entomol.* **2002**, *53*, 323–343. [CrossRef]
38. Aenishaenslin, C.; Bouchard, C.; Koffi, J.K.; Ogden, N.H. Exposure and preventive behaviours toward ticks and Lyme disease in Canada: Results from a first national survey. *Ticks Tick-Borne Dis.* **2017**, *8*, 112–118. [CrossRef] [PubMed]
39. Beaujean, D.J.M.A.; Bults, M.; van Steenbergen, J.E.; Voeten, H.A.C.M. Study on public perceptions and protective behaviors regarding Lyme disease among the general public in the Netherlands: Implications for prevention programs. *BMC Public Health* **2013**, *13*, 225. [CrossRef] [PubMed]
40. Belova, O.A.; Burenkova, L.A.; Karganova, G.G. Different tick-borne encephalitis virus (TBEV) prevalences in unfed versus partially engorged ixodid ticks-Evidence of virus replication and changes in tick behavior. *Ticks Tick-Borne Dis.* **2012**, *3*, 240–246. [CrossRef]
41. Schulze, T.L.; Stafford, K.C., III; Heffernan, R.; Fish, D.; des Vignes, F.; Piesman, J. Effect of Tick Removal on Transmission of Borrelia burgdorferi and Ehrlichia phagocytophila by Ixodes scapularis Nymphs. *J. Infect. Dis.* **2002**, *183*, 773–778. [CrossRef]
42. Heiervang, E.; Goodman, R. Advantages and limitations of web-based surveys: Evidence from a child mental health survey. *Soc. Psychiaty Psychiatr. Epidemiol.* **2011**, *46*, 69–76. [CrossRef]
43. Huang, Y.; Xu, S.; Lei, W.; Zhao, Y.; Liu, H.; Yao, D.; Xu, Y.; Lv, Q.; Hao, G.; Xu, Y.; et al. Knowledge, attitudes, and practices regarding zika: Paper and Internet Based Survey in Zhejiang, China. *JMIR Public Health Surveill.* **2017**, *3*, 81. [CrossRef]
44. Betsch, C.; Schmid, P.; Heinemeier, D.; Korn, L.; Holtmann, C.; Böhm, R. Beyond confidence: Development of a measure assessing the 5C psychological antecedents of vaccination. *PLoS ONE* **2018**, *13*, e0208601. [CrossRef] [PubMed]

45. Moßhammer, D.; Michaelis, M.; Mehne, J.; Wilm, S.; Rieger, M.A. General practitioners' and occupational health physicians' views on their cooperation: A cross-sectional postal survey. *Int. Arch. Occup. Environ. Health* **2016**, *89*, 449–459. [CrossRef] [PubMed]
46. Alicino, C.; Iudici, R.; Barberis, I.; Paganino, C.; Cacciani, R.; Zacconi, M.; Battistini, A.; Bellina, D.; Bella, A.M.D.; Talamini, A.; et al. Influenza vaccination among healthcare workers in Italy: The experience of a large tertiary acute-care teaching hospital. *Hum. Vaccines Immunother.* **2015**, *11*. [CrossRef] [PubMed]

Tropical Medicine and Infectious Disease

Article

Vertical Infestation Profile of *Aedes* in Selected Urban High-Rise Residences in Malaysia

Nurulhusna Ab Hamid [1,*], Siti Nurfadhlina Mohd Noor [1], Nur Rasyidah Isa [1], Rohaiyu Md Rodzay [1], Ainaa Mardia Bachtiar Effendi [1], Afiq Ahnaf Hafisool [1], Fatin Atirah Azman [1], Siti Farah Abdullah [1], Muhammad Khairi Kamarul Zaman [1], Mohd Iqbal Mohd Norsham [1], Noor Hasmiza Amanzuri [1], Nurliyana Abd Khalil [1], Izzah Farhah Zambari [1], Aimannur Najihah Mat Rani [1], Farah Diana Ariffin [1], Topek Omar [2], Nazni Wasi Ahmad [1] and Han Lim Lee [1]

1 Medical Entomology Unit, WHO Collaborating Centre for Vectors, Institute for Medical Research, Ministry of Health, Malaysia, National Institutes of Health, Block C, Jalan Setia Murni U13/52, Seksyen U13, Setia Alam, Shah Alam 40170, Malaysia; snfadhlina@gmail.com (S.N.M.N.); nurrasyidahisa@gmail.com (N.R.I.); rohaiyuRodzay@gmail.com (R.M.R.); ainaaeffendi@gmail.com (A.M.B.E.); afiqahnaf94@gmail.com (A.A.H.); fatinazman94@gmail.com (F.A.A.); abdullahsitifarah@gmail.com (S.F.A.); eryian30@yahoo.com (M.K.K.Z.); iqbal_sham@live.com (M.I.M.N.); osakurataxo@yahoo.com (N.H.A.); liyana_khalil@yahoo.com (N.A.K.); izzahzambari@gmail.com (I.F.Z.); anmatrani@gmail.com (A.N.M.R.); farahdiana.a@moh.gov.my (F.D.A.); nazni@moh.gov.my (N.W.A.); mosquito7090@gmail.com (H.L.L.)

2 Entomology and Pest Unit, Federal Territory of Kuala Lumpur & Putrajaya Health Department, Jalan Cenderasari, Kuala Lumpur 50590, Malaysia; aks.topek@gmail.com

* Correspondence: husna.hamid@moh.gov.my

Received: 8 May 2020; Accepted: 13 June 2020; Published: 7 July 2020

Abstract: Dengue is placing huge burdens on the Malaysian healthcare system as well as the economy. With the expansion in the number of high-rise residential buildings, particularly in the urban centers, the flight range and behavior of *Aedes* mosquitoes may be altered in this habitat type. In this study, we aimed to expand the understanding of the vertical distribution and dispersal of *Aedes* in nine selected high-rise residences in Kuala Lumpur, Selangor, and Johor using ovitraps as the sampling method. We discovered that *Ae. aegypti* is the predominant species in all study sites. Both *Ae. aegypti* and *Ae. albopictus* are most abundant within the first three levels and could be found up to level 21 (approximately 61.1–63.0 m). Pearson correlation analyses exhibited negative correlations in eight out of nine study sites between the ovitrap indexes (OIs) within each floor level, suggesting that *Aedes* density decreased as the building level increased. Our findings provide information to the public health authorities on 'hot spot' floors for effective suppression of dengue transmission.

Keywords: vertical dispersal; *Aedes*; mosquito; high-rise residences; dengue; surveillance

---

1. Introduction

Dengue is a vector-borne infectious disease with an estimated 100–400 million infections annually [1]. Transmission of one of the four antigenically distinct serotypes of dengue virus (DENV-1 to DENV-4) may cause dengue and its severe forms—hemorrhagic fever and shock syndrome [2]. To date, efficacious and cost-effective vaccines and antiviral drugs against these four serotypes are still under various stages of research and development [3,4]. Until dengue vaccines or antiviral drugs become available, the only current method proven to be effective in dengue control and prevention is vector control measures that could be sustained through community involvement [1]. The spread of dengue globally is fueled by a combination of various factors including climate change, rapid urbanization, increased international travel and trade, as well as changes in land-use patterns [5].

Dengue is endemic in Malaysia and Malaysia is one of the most severely affected countries in South-East Asia [1]. The country had 80,615 reported dengue cases and 147 deaths in the year 2018 [6]. Selangor reported the highest number of dengue cases (45,349 cases), followed by Federal Territories of Kuala Lumpur and Putrajaya (7591 cases), Penang (6071 cases) and Johor (5885 cases) [6]. The primary mosquito vector for dengue is *Aedes aegypti*, while *Ae. albopictus* is recognized as the secondary vector [2]. These two *Aedes* species are prevalent in Malaysia where they can coexist in similar ecological niches [7].

Increased demand for housing due to urban sprawl and land scarcity in major urban regions resulted in an increase in high-rise residential buildings ranging from flats and apartments to luxury condominiums [8]. According to a housing statistic, 5.5 million people in Malaysia live in high-rise residences as of June 2014 [9], suggesting that a large number of Malaysians are experiencing high-rise living, also known as vertical living. The rise in the numbers of new non-landed, high-rise properties, as a result, may pose greater challenges in developing preventive and control measures against dengue, due to a change in dispersal dynamics of *Aedes* due to adaptation to non-landed, high-rise housing.

The vector flight range plays a key role in the dynamics of transmission. In standard approaches by the Ministry of Health (MOH), Malaysia, reduction of the proliferation of infected vectors and suppression of dengue outbreaks are achieved through vector control measures, e.g., insecticide fogging, larviciding and elimination of potential larval habitats, within a buffer zone with a radius of 200 m from the house where a dengue case is reported [10]. This is in line with standard mark-release-recapture studies showing that the mean distance of recaptured *Ae. aegypti* rarely went beyond 100 m [11,12]. In a review article, *Ae. aegypti* is said to disperse horizontally with an average distance of 83.4 m [13]. However, the standard approaches used are generally applied to low-rise housing. Slightly different vector control approaches may be required for high-rise housing. At present, specific guidelines for vector control activities in high-rises are yet to be released. But according to a circular from the MOH, fogging shall be performed within a 200 m radius from the index case house, and for buildings that have more than five levels, fogging shall be completed at least within five levels below and above the index case house [14]. For buildings that have five floors or less, the whole building shall be fogged [14].

Studies on the vertical distribution of *Aedes* in high-rises have been carried out in several countries, including Sri Lanka [15], Trinidad [16], Singapore [17] and Malaysia [7,18,19]. Research in Trinidad, West Indies, suggested that *Aedes* eggs were found mostly in 13–24 m elevations [16]. While in Sri Lanka, *Aedes* could be found from the ground floor to the highest floor of 130 ft (approximately 40 m), but largely were discovered at the 60 ft elevation (approximately 18 m) [15]. A similar observation was seen in Putrajaya, Malaysia in which *Aedes* density peaked at the height of 16.1–18 m [19]. These studies indicated that infestations of *Aedes* in high-rises were extensive, ranging from the ground level to higher elevations, and that they have the ability to survive in high-rise buildings.

This study aimed to expand the understanding of the vertical distribution of *Aedes* in nine selected high-rise residences in the Federal Territory of Kuala Lumpur and the States of Selangor and Johor using ovitraps as the sampling method. Although a few studies have been performed in Malaysia for similar purposes [7,18–20], the ever-increasing number of high-rise residences in Malaysia makes this study warranted. Up-to-date information gained from this study will hopefully serve as guidelines for the public health officers to identify 'hot spot' floors of *Aedes* infestation in high-rises for quick and more efficient dengue control efforts.

2. Materials and Methods

2.1. Study Sites

The study was conducted in nine study sites that constituted high-rise properties. Five study sites were in Selangor State—Subang Perdana Goodyear Court 8 (GC8), Goodyear Court 10 (GC10), Apartmen Pesona (AP), Flat Sri Kota (FSK) and PPR Taman Mulia (PTM); two study sites were in Johor State—PPR Taman Kempas Permai (TKP) and Apartmen Sri Wangi (ASW) and two study sites were in the Federal Territory of Kuala Lumpur—Kuarters Jalan Sultan Abdul Aziz Block A (KJA) and Block C (KJC) (Figure 1). The sites were chosen because they had prolonged dengue outbreaks between the years 2014 and 2018. All study sites were in urban areas, at which their coordinates were obtained using the Global Positioning System (GPS) (Garmin Montana®680, Olathe, KS, USA). The physical and geographical description as well as the building design of each study site are explained in Supplementary Table S1 and Figures S1–S6.

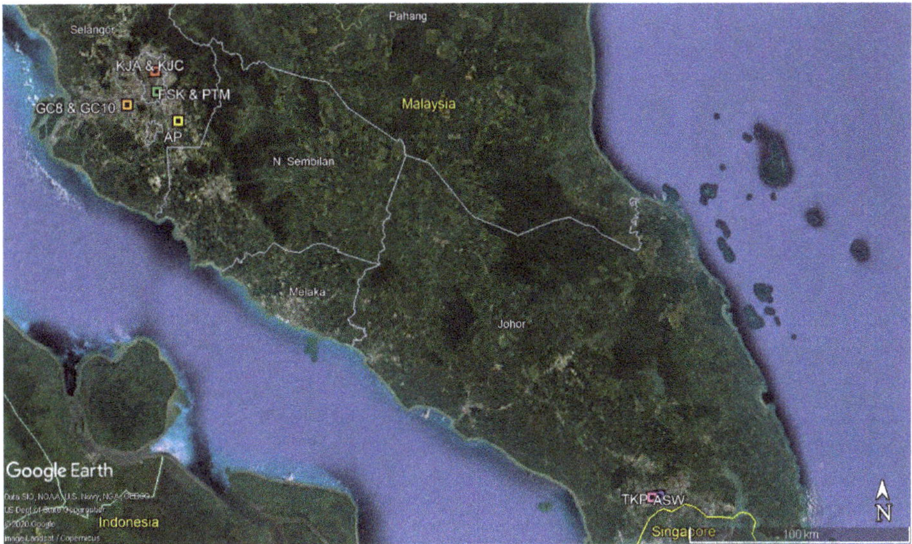

Figure 1. The geographical map of the study sites. Abbreviations: GC8—Goodyear Court 8, GC10—Goodyear Court 10, AP—Apartmen Pesona, FSK—Flat Sri Kota, PTM—PPR Taman Mulia, KJA—Kuarters Jalan Sultan Abdul Aziz Block A, KJC—Kuarters Jalan Sultan Abdul Aziz, TKP—PPR Taman Kempas Permai and ASW—Apartmen Sri Wangi.

2.2. Ovitrap Surveillance

The number of *Aedes* larvae that represented the population densities of *Aedes* was assessed at different building levels via ovitraps. Ovitrapping was performed as described in Lee [21]. Each ovitrap consisted of a 250 mL cylindrical, black plastic container (7.0 cm diameter, 9.0 cm height) filled with tap water to a level of 5.5 cm. Each ovitrap was equipped with a removable oviposition paddle that was made from a thin strip of brown hardboard (10 cm × 2.5 cm × 0.3 cm). The paddle was placed diagonally with the rough surface facing upwards where the mosquitoes laid eggs above the water level.

Ovitrap monitoring was conducted separately for each study site between the years 2014 and 2018, with sampling at each site consisting of 10 weeks of data collection. The total number of ovitraps placed in each study site is described in Table 1. One ovitrap was placed randomly on each level that had minimum human, physical and environmental disturbance. We placed the ovitraps in a semi-indoor environment, defined as the area outside of the house units but is still sheltered by the roof, e.g., shared corridor and stairway. The ovitraps were collected after seven days and replaced with new ovitraps consisting of fresh tap water and egg-free oviposition paddles. Any disturbances to the exposed ovitraps, e.g., theft, vandalism or invasion by insects, were recorded.

The collected ovitraps were transported back to the Institute for Medical Research (IMR), Kuala Lumpur, laboratory for further processing. The ovitrap contents along with the oviposition paddle were transferred into the plastic containers and labeled according to the study site, date of collection, and level. Beef liver powder (Difco Laboratories, MD, USA) was added into each container as larval food. Identification to species level of third or fourth instar was performed using established taxonomy keys [22,23] under a compound microscope (Nikon Eclipse® E100, Japan). The ovitrap contents were examined for species identification until no newly emerged larvae were present in the containers.

Table 1. The total number of ovitraps placed in each study site.

Study Site	Total Number of Ovitraps
GC8	300
GC10	500
AP	200
FSK	700
PTM	400
ASW	360
TKP	510
KJA	570
KJC	190

2.3. Data Analysis

Data entry and statistical analyses were conducted using Microsoft Excel and Statistical Package for Social Sciences (SPSS) Version 25.0 [24]. Ovitrap index (OI) is expressed as the percentage of positive ovitraps per the number of ovitraps recovered. Summation and the mean number of *Aedes* larvae for each *Aedes* species per ovitrap were performed in every study site. Paired samples *t*-test analysis was conducted to determine if there is any significant difference between the mean number of *Ae. aegypti* larvae and *Ae. albopictus* larvae per ovitrap. Single and cohabitation of *Aedes* spp. were also analyzed. Linear regression line for OIs in each floor level was constructed for each site. The Spearman's rank correlation coefficient test was further conducted to investigate the correlation between the mean number of *Aedes* larvae and floor level. All levels of statistical significance were determined at $p \leq 0.05$.

3. Results

The overall results showed that the OI varied quite considerably between the nine study sites, with AP exhibiting the highest OI (63.00 ± 3.40%), followed by TKP (55.00 ± 2.40%) and ASW (51.00 ± 2.70%) (Table 2). The lowest OI belonged to KJA with OI of 26.00 ± 1.90% (Table 2). Paired samples *t*-test analysis indicated that the mean number of *Ae. aegypti* larvae per ovitrap was significantly higher than that of *Ae. albopictus* in all study sites ($p \leq 0.05$), demonstrating that *Ae. aegypti* is the predominant species in all sites studied. Collected *Ae. aegypti* larval instars were highest in TKP (8847 larvae), while *Ae. albopictus* instars were most abundant in AP (980 larvae) (Table 2). The relative abundance (ratio) of *Ae. aegypti* to *Ae. albopictus* ranged from the lowest of 1.7:1 in AP to the highest of 30.6:1 in KJA (Table 2).

Table 2. The overall analysis of *Aedes* larval instars collected in each study site.

Study Site	Ovitrap Collected	Ovitrap Index (%)	*Ae. aegypti*		*Ae. albopictus*		Ratio *Ae. aegypti*: *Ae. albopictus*
			Total Larvae	Overall Percentage (%)	Total Larvae	Overall Percentage (%)	
GC8	297/300	48.00 ± 2.90	5216	94.02	332	5.98	15.7: 1
GC10	454/500	41.00 ± 2.30	5258	84.63	955	15.37	5.5: 1
AP	200/200	63.00 ± 3.40	1640	62.60	980	37.40	1.7: 1
FSK	658/700	37.00 ± 1.90	3111	91.10	304	8.90	10.3: 1
PTM	371/400	37.00 ± 2.50	1614	76.17	505	23.83	3.2: 1
ASW	351/360	51.00 ± 2.70	4483	90.22	486	9.78	9.3: 1
TKP	448/510	55.00 ± 2.40	8847	92.45	722	7.55	12.3: 1
KJA	564/570	26.00 ± 1.90	1898	96.79	63	3.21	30.6: 1
KJC	188/190	31.00 ± 3.40	738	94.98	39	5.02	18.7: 1

We performed a further analysis by looking at single and cohabitation of *Aedes* species in the same ovitraps. Our analyses showed that single breeding of *Aedes* was much more prevalent than the cohabitation of *Aedes* spp. in all study sites (Table 3). The proportions of ovitraps positive for single breeding of *Ae. aegypti* were noticeably high in all study sites, ranging from 64.80% in AP to 96.62% in KJA (Table 3). It is worth noting that KJA was the only study site void of cohabitation of *Aedes* spp., contributing to its high percentage of single breeding of *Ae. aegypti* (Table 3). Meanwhile, percentages of ovitraps positive for single breeding of *Ae. albopictus* were, in general, higher than cohabitation of *Aedes* spp., with the exception of TKP (Table 3). Analysis variance of one-way ANOVA showed that there were significant differences in the mean number of *Ae. aegypti* larvae ($p \leq 0.05$, F = 34.652) and *Ae. albopictus* larvae ($p \leq 0.05$, F = 9.712) in single breeding as well as in cohabitation of *Aedes* spp. among the nine study sites ($p \leq 0.05$, F = 8.628). TKP displayed the highest mean number of *Ae. aegypti* per ovitrap in single breeding (17.33 ± 1.50) and cohabitation of *Aedes* spp. (3.58 ± 0.74) (Table 3). Interestingly, PTM was the only site that exhibited a higher ratio of *Ae. albopictus* to *Ae. aegypti* in cohabitation containers (Table 3).

Table 3. Comparisons between single breeding and cohabitation of *Aedes* spp. in each study site.

Study Site	Single Breeding				Cohabitation		
	Ae. aegypti		*Ae. albopictus*				
	Percentage of Positive Ovitraps (%)	Mean Larvae per Ovitrap ± SE	Percentage of Positive Ovitraps (%)	Mean Larvae per Ovitrap± SE	Percentage of Positive Ovitraps (%)	Mean Larvae per Ovitrap± SE	Ratio *Ae. aegypti*: *Ae. albopictus*
GC8	90.97	16.96 ± 1.66	5.56	0.79 ± 0.37	3.47	0.93 ± 0.45	1.86: 1.00
GC10	78.38	10.56 ± 0.99	13.51	1.40 ± 0.39	8.11	1.73 ± 0.57	1.45: 1.00
AP	64.80	7.38 ± 0.91	21.60	4.14 ± 1.18	13.60	1.59 ± 0.45	1.08: 1.00
FSK	90.24	4.55 ± 0.34	6.91	0.38 ± 0.11	2.85	0.26 ± 0.10	2.31: 1.00
PTM	78.68	4.25 ± 0.50	18.38	1.16 ± 0.31	2.94	0.30 ± 0.16	1.00: 1.90
ASW	79.33	11.32 ± 1.22	10.61	0.89 ± 0.27	10.61	1.95 ± 0.51	2.10: 1.00
TKP	82.26	17.33 ± 1.50	6.05	0.45 ± 0.16	11.69	3.58 ± 0.74	2.92: 1.00
KJA	96.62	3.37 ± 0.30	3.38	0.11 ± 0.07	0.00	0.00 ± 0.00	0.00: 0.00
KJC	94.92	3.76 ± 0.53	3.39	0.11 ± 0.09	1.69	0.27 ± 0.27	1.78: 1.00

The relationship between OI and floor level was investigated in this study. For this, the linear regression line of each study site was constructed (Figure 2). Except for KJA, Pearson correlations exhibited negative correlations in all sites, suggesting that as the floor level increased, the OI decreased. Significant differences between OIs within each floor level were exhibited in GC8 ($p = 0.020$), GC10 ($p = 0.010$), FSK ($p = 0.009$) and PTM ($p = 0.007$). KJA had a positive correlation between OIs within each floor level but was insignificant ($p = 0.460$).

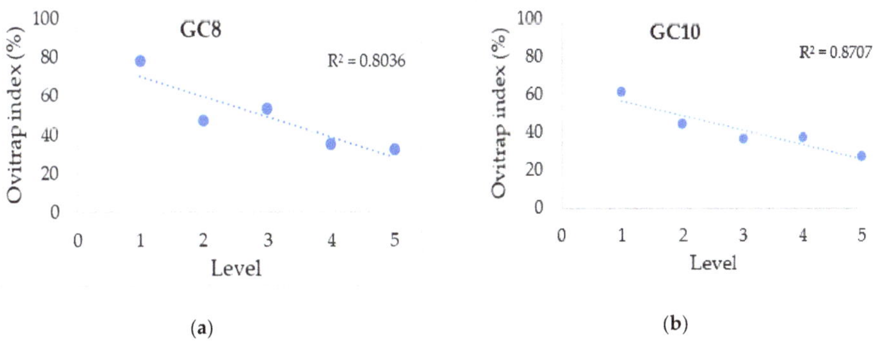

(a) (b)

Figure 2. *Cont.*

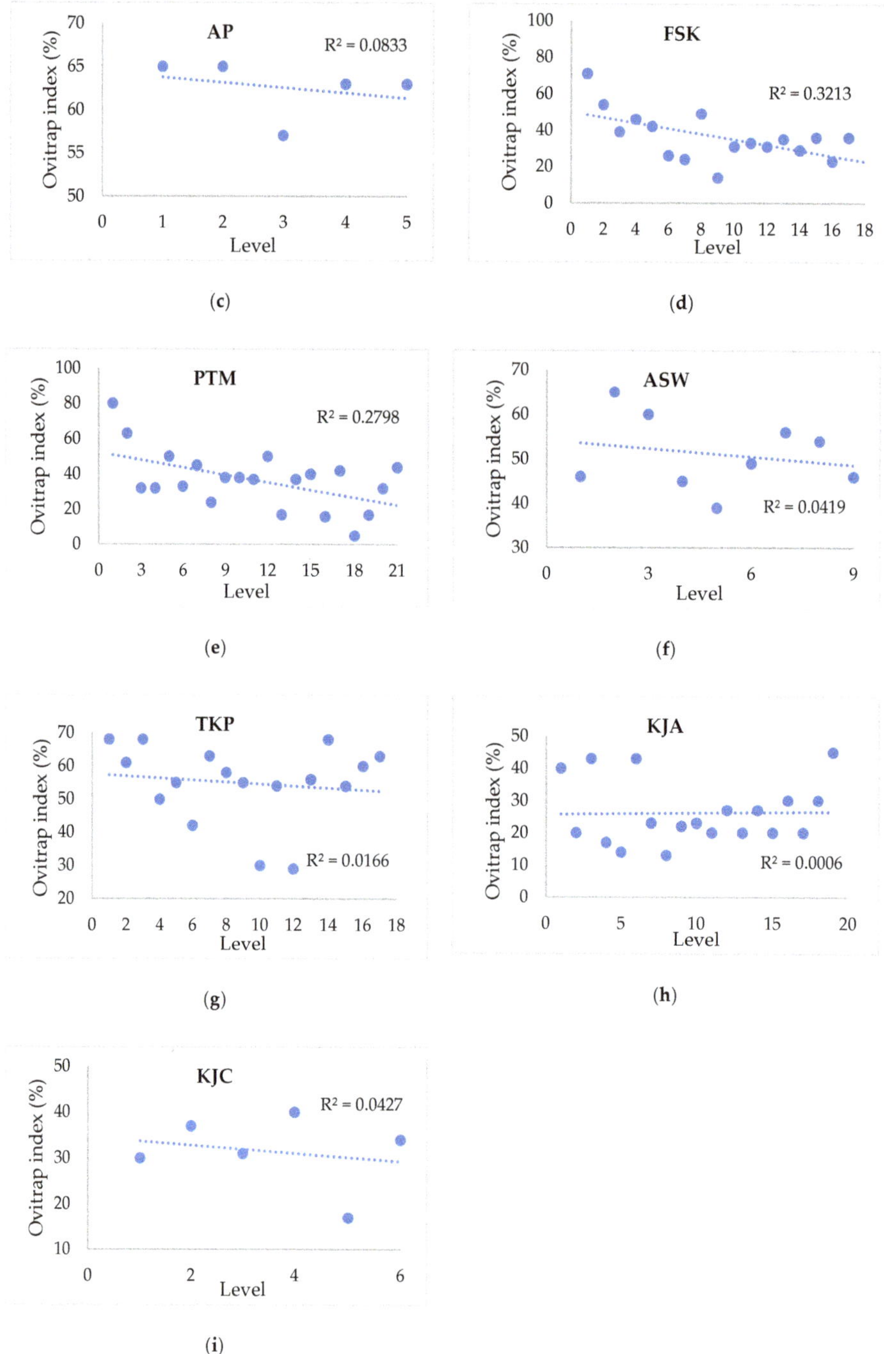

Figure 2. Relationship between ovitrap index (%) and level in each study site. (**a–i**) GC8, GC10, AP, FSK, PTM, ASW, TKP, KJA, KJC.

To go deeper into this analysis, we found that six out of nine study sites—GC8, GC10, AP, FSK, PTM, and TKP—displayed the highest mean number of *Aedes* larvae per ovitrap on floor level 1 (0.0–3.0 m height) (Table 4). In contrast, *Aedes* larvae per ovitrap peaked on the top floor (54.1–57.0 m height) in KJA (Table 4). In ASW and KJC, *Aedes* larvae density was highest on level 7 (18.1–21.0 m height) and level 4 (9.1–12 m height), respectively (Table 4). Intriguingly, all three sites of the five-story building—GC8, GC10 and AP—showed the lowest mean number of *Aedes* larvae per ovitrap on the highest floor (12.1–15.0 m height) (Table 4). Likewise, ASW, KJA and KJC presented the least *Aedes* larvae per ovitrap within the same floor height (Table 4). It may be noted that *Ae. aegypti* was found on every floor level at all study sites, up to level 21 (61.1–63.0 m height). Similar to *Ae. aegypti*, *Ae. albopictus* was discovered on each floor level up to level 21, but in KJA and KJC, the species could only be found up to level 3 and level 2, respectively (Supplementary Tables S2 and S3). Similar to the OI analysis, Spearman's rank correlation coefficient showed negative correlations in all sites except for KJA (Table 4).

Table 4. The mean number of *Aedes* larvae per ovitrap on each level. Spearman's rank correlation coefficient between the mean number of *Aedes* larvae and level (height) was calculated using SPSS statistics.

Level	Approximate Height (m)	GC8	GC10	AP	FSK	PTM	ASW	TKP	KJA	KJC
1	0.0–3.0	36.16 ± 5.02	26.72 ± 3.71	14.90 ± 2.80	9.88 ± 1.40	14.10 ± 3.40	11.21 ± 3.08	36.29 ± 8.81	5.13 ± 1.73	4.10 ± 1.48
2	3.1–6.0	19.28 ± 3.91	12.99 ± 2.07	13.28 ± 2.68	6.57 ± 1.66	9.16 ± 2.62	15.83 ± 3.68	33.39 ± 8.59	1.67 ± 0.77	4.93 ± 1.53
3	6.1–9.0	14.22 ± 2.76	10.73 ± 2.25	14.18 ± 3.68	3.63 ± 1.05	6.05 ± 2.27	17.38 ± 3.82	20.21 ± 4.00	3.73 ± 1.07	3.76 ± 1.20
4	9.1–12.0	13.72 ± 3.34	10.04 ± 1.97	12.08 ± 3.99	6.23 ± 1.60	7.58 ± 3.56	9.38 ± 2.68	23.42 ± 7.21	2.59 ± 1.13	5.47 ± 1.52
5	12.1–15.0	10.53 ± 2.73	8.76 ± 2.08	11.08 ± 2.19	5.87 ± 1.42	7.83 ± 2.90	7.89 ± 2.67	17.90 ± 4.86	1.31 ± 0.91	2.23 ± 1.11
6	15.1–18.0				4.64 ± 1.59	5.44 ± 3.49	15.64 ± 4.04	14.75 ± 5.71	4.67 ± 1.12	4.31 ± 1.71
7	18.1–21.0				4.92 ± 1.81	5.45 ± 2.31	19.79 ± 5.25	25.25 ± 7.09	4.30 ± 1.75	
8	21.1–24.0				8.38 ± 1.81	3.12 ± 1.79	15.64 ± 4.41	17.21 ± 5.65	2.27 ± 1.28	
9	24.1–27.0				2.76 ± 1.51	5.50 ± 2.13	14.51 ± 4.21	12.86 ± 4.38	2.22 ± 0.97	
10	27.1–30.0				4.08 ± 1.28	3.25 ± 1.93		16.55 ± 8.33	4.07 ± 1.59	
11	30.1–33.0				1.53 ± 1.48	8.11 ± 3.44		19.61 ± 7.04	2.73 ± 1.09	
12	33.1–36.0				3.72 ± 1.29	9.06 ± 2.94		5.54 ± 2.86	3.03 ± 1.01	
13	36.1–39.0				3.15 ± 1.03	1.28 ± 0.80		23.00 ± 6.57	4.00 ± 1.68	
14	39.1–42.0				4.00 ± 1.41	4.58 ± 1.45		33.68 ± 7.43	2.80 ± 1.01	
15	42.1–45.0				5.23 ± 1.49	5.10 ± 2.69		10.86 ± 2.93	2.90 ± 1.23	
16	45.1–48.0				3.08 ± 1.10	0.89 ± 0.57		32.48 ± 8.58	2.97 ± 1.09	
17	48.1–51.0				5.31 ± 1.71	6.68 ± 2.46		17.73 ± 5.33	3.07 ± 1.18	
18	51.1–54.0					0.05 ± 0.05			3.90 ± 1.33	
19	54.1–57.0					1.17 ± 0.76			8.66 ± 2.01	
20	57.1–60.0					6.00 ± 2.29				
21	60.1–63.0					7.28 ± 4.52				
r		−0.304	−0.231	−0.070	−0.172	−0.203	−0.024	−0.059	0.024	−0.028
p		0.000	0.000	0.322	0.000	0.000	0.657	0.211	0.576	0.705

4. Discussion

With the expansion in the number of high-rise residential buildings largely in the urban areas in Malaysia, we believe that the up-to-date information on the vertical distribution of *Aedes* in this habitat type is warranted. We discovered that *Ae. aegypti* was the predominant species in all nine study sites, in which the number of *Ae. aegypti* larval instars were up to 30-fold higher in comparison to *Ae. albopictus*. The dominance of *Ae. aegypti* is in line with the results of several vertical dispersal studies conducted in urban high-rise residences in Kuala Lumpur and Selangor [7,18] as well as Putrajaya [19]. This finding is not surprising, because *Ae. aegypti* is adapted to reside in and around human dwellings. Female *Ae. aegypti* also preferentially breed in artificial and domestic containers in the urban areas [25], which may be commonly found in the semi-indoor environment of our study sites. Likewise, the lower number of *Ae. albopictus* larval instars observed in all study sites were due to their behavior. *Ae. albopictus* prefers to breed in natural habitats such as tree holes and to reside where vegetation is plentiful [26], which may be especially uncommon in high-rise properties where space is limited for planting or gardening. However, recent studies in Malaysia discovered that *Ae. albopictus* have been acclimatizing to the domestic environment in urban areas. The species is now found both in the indoor and outdoor environments and readily breeds in artificial or man-made containers [27,28]. Therefore, it is not surprising that *Ae. albopictus* was found, but in a lower number.

The mixed infestation of *Ae. aegypti* and *Ae. albopictus* was detected in all sites, apart from KJA, accounting for 1.39% to 13.60% of the percentage of positive ovitraps collected. The cohabitation of *Aedes* spp. found in the present study suggested that their ecological niche overlapped. A study in multi-story buildings in Selangor and Kuala Lumpur by Lau [7] discovered a higher percentage of cohabitation of *Aedes* spp., ranging from 10.77% to 26.56% from the total percentage of positive ovitraps collected. The lower percentages of co-breeding compared to single breeding may be explained through their breeding behavior. *Ae. albopictus*, for instance, would avoid breeding in habitats occupied by *Ae. aegypti*, and vice versa, as shown in a study in Penang [29].

The discovery of *Aedes* on each level suggested that *Aedes* could disperse up to the highest floor level. Three main types of adult mosquito dispersal exist [30]. The first is unintentional dispersal by riding along with humans via airplanes or ships, involving long-distance dispersal from one continent to another. The second is mosquitoes being carried through wind-assisted dispersal, also in a long-distance and unintentional dispersal. The third type (involved in our study) is considered as active and intentional in a short-distance dispersal. In high-rise buildings, *Aedes* could be involved in short-distance dispersal via human transportation using stairs or lifts as suggested in other studies [7,19]. Mosquitoes may disperse to look for blood sources, nectar sources, mates, oviposition sites and resting sites [30]. The search for new oviposition sites, for example, could be represented by the ovitraps and possibly other cryptic breeding sites that drive the dispersal of female *Aedes* through the availability of stagnant water for the development of immature, aquatic stages of mosquitoes. Studies conducted in urban habitats in Malaysia discovered that both *Aedes* spp. were mostly found in domestic receptacles such as flower vases, flower pot plates, pails, bowls, refrigerator trays, plastic containers, empty paint cans and in building structures such as roofs, drains, gutters and gully traps [27,31,32]. We hypothesized that the *Aedes* mosquitoes in our study sites could have originated from these sources.

Additionally, the discovery of *Aedes* on each level suggested that the ground floor, as well as the higher elevations, support the survival of *Aedes*. The availability of biotic and abiotic components (ecological niche) and ecosystem in our study sites may have provided sufficient bloodmeal, resting place, and breeding sites for *Aedes* [16]. Biotic components include humans, pets, and plants, while abiotic components include temperature, humidity, wind speed, and building structure [16]. The denser people living in high-rises will most likely provide more blood meals for *Aedes*, and in turn increase their vectorial capacity to transmit dengue viruses due to higher contact with human hosts [33]. This is particularly true for *Ae. aegypti*, which has an anthropophilic tendency for preferential feeding on human blood and takes multiple blood-feeding in a gonotrophic cycle [34,35].

Although *Aedes* dispersal seemed extensive (up to level 21, 61.0–63.0 m height), the present study demonstrated that *Aedes* larvae were most abundant on the ground floor. A plausible reason may be due to the existence of communal areas in the ground level such as parking spaces, shops and waste disposal areas that sometimes could have poor sanitation, creating favorable mosquito breeding sites. In Singapore, for example, children residing in the ground level reportedly had higher dengue infection rates compared to those residing in higher levels because more *Aedes* breeding habitats were found in the communal areas [17]. This finding emphasized that the ground floor demands additional efforts during vector control measures.

Intriguingly, unlike the other eight other study sites, we found *Aedes* larvae were most abundant at the highest level of KJA with an elevation of approximately 54.1 to 57 m. We speculate that this might be due to the presence of an uneven rooftop floor, where heavy rain will cause puddles and create breeding sites for mosquitoes. Importantly, in the study, we found that the maximum height of *Ae. aegypti* dispersal was identical to *Ae. albopictus* (61.1–63.0 m). However, we wanted to point out that *Ae. albopictus* had overall a higher density closer to the ground floor. In KJA and KJC, they could only be found up to level 2 and level 3, respectively. The results suggested that *Ae. albopictus* preferred elevations closer to the ground floor, presumably with scattered vegetation around for breeding and resting. Furthermore, from the results, it was clear that 'hot spot' floors of *Aedes* infestation were within the first three floors. Again, we think that floors closer to the ground may create a complete ecosystem and ecological niche suitable for *Aedes* infestation.

It is noteworthy to point out that TKP exhibited comparably high *Aedes* populations at the upper levels (levels 14 to 17) as well as the first two floor levels, implying a U-shaped curve instead of a declining trend. This may have occurred owing to the existence of water tanks between levels 16 and 17. Likewise, PTM has water tanks between levels 20 and 21 that may be responsible for the considerably large *Aedes* populations at the upper levels.

Our study possesses some limitations. Firstly, we did not consider environmental factors such as temperature, relative humidity, wind speed and rainfall. Wind speed, for instance, could influence the flight capabilities of mosquitoes [36], while relative humidity influences the mating and feeding behavior of mosquitoes [37]. Future research incorporating these environmental factors may be warranted. Secondly, our study sites have different building designs that may have influenced our results. However, we could say that the low-cost houses, i.e., AP, FSK, TKP, and PTM, possess similar building designs (Supplementary Figures S2, S3 and S5). Because our study only included low- and medium-cost houses, future studies incorporating high-cost houses such as luxury condominiums may provide interesting new insights into the vertical dispersal of *Aedes*.

In conclusion, our study revealed that the dispersal of *Aedes* vertically in high-rise residential buildings could be considered extensive. We think that *Aedes* could disperse to the upper floors even under little pressure to fulfill their needs, particularly finding hosts and oviposition sites. Based on our results, we recommended that the vector control operations should be concentrated within the first three floors and that guidelines specific for high-rises should be available. Given the extensive vertical dispersal of *Aedes*, other levels (middle and upper) should also be taken into consideration during vector control measures, depending on the location of the index case house as recommended in the circular. Vertical vector dispersal and propagation may play key roles in dengue virus dissemination in high-rises and may require further attention to fully investigate its importance.

Supplementary Materials: The following are available online at http://www.mdpi.com/2414-6366/5/3/114/s1, Table S1: Physical and geographical descriptions of all study sites. Figure S1–S6: Building designs of all study sites. Table S2: Total number and percentage of *Aedes* spp. on each level at KJA. Table S3: Total number and percentage of *Aedes* spp. on each level at KJC.

Author Contributions: Conceptualization, N.A.H., S.N.M.N., N.W.A., H.L.L.; methodology, N.A.H., H.L.L., T.O., N.W.A.; software, N.R.I., A.A.H., M.K.K.Z.; validation, N.A.H., F.D.A., T.O., N.W.A., H.L.L.; formal analysis, N.A.H., S.N.M.N., N.R.I.; investigation: A.M.B.E., A.A.H., F.A.A., S.F.A., M.K.K.Z., M.I.M.N., N.H.A., N.A.K., I.F.Z.; A.N.M.R.; resources: N.A.H., F.D.A., T.O., N.W.A., H.L.L.; data curation: N.A.H., S.N.M.N., N.R.I., R.M.R.; writing—original draft preparation, N.A.H., S.N.M.N., R.M.R.; writing—review and editing: N.A.H., F.D.A., T.O., H.L.L.; supervision, N.A.H., F.D.A., H.L.L.; project administration, N.A.H., F.D.A., T.O., N.W.A., H.L.L. All authors have read and agreed to the published version of the manuscript.

Funding: This research was funded by the Ministry of Health Malaysia Research Grant (NMRR-13-921-17915).

Acknowledgments: The authors thank the Director-General of Health, Malaysia for his permission to publish this article, and Director of Institute for Medical Research for unceasing support. The authors also thank the staff of the Medical Entomology Unit, IMR for assistance in the field.

Conflicts of Interest: The authors declare no conflict of interest.

Ethical Approval: The study was approved by the Institute for Medical Research Committee (JPP-IMR) and National Institutes of Health Malaysia (JPP-NIH), and registered with the National Medical Research Register (NMRR-13-921-17915). This study met the ethical standard of the Medical Research and Ethics Committee (MREC) (Ref No: KKM/NIHSEC/800–2/2/2JldP13905).

References

1. WHO. Dengue and Severe Dengue. Available online: https://www.who.int/news-room/fact-sheets/detail/dengue-and-severe-dengue (accessed on 15 April 2020).
2. WHO. *Dengue Guidelines for Diagnosis, Treatment, Prevention and Control: New Edition*; World Health Organization: Geneva, Switzerland, 2009.

3. Deng, S.Q.; Yang, X.; Wei, Y.; Chen, J.T.; Wang, X.J.; Peng, H.J. A review on dengue vaccine development. *Vaccines* **2020**, *8*, 63.
4. Kato, F.; Ishida, Y.; Oishi, S.; Fujii, N.; Watanabe, S.; Vasudevan, S.G.; Tajima, S.; Takasaki, T.; Suzuki, Y.; Ichiyama, K.; et al. Novel antiviral activity of bromocriptine against dengue virus replication. *Antivir. Res.* **2016**, *131*, 141–147.
5. WHO. WHO Region of the AMERICAS Records Highest Number of Dengue Cases in History; Cases Spike in Other Regions. Available online: https://www.who.int/news-room/detail/21-11-2019-who-region-of-the-americas-records-highest-number-of-dengue-cases-in-history-cases-spike-in-other-regions (accessed on 15 April 2020).
6. Ministry of Health Malaysia. Kenyataan Akhbar Ketua Pengarah Kesihatan Malaysia. Situasi semasa Demam Denggi, Zika, dan Chikungunya di Malaysia. Available online: http://www.moh.gov.my/index.php/database_stores/store_view_page/17/1119 (accessed on 15 April 2020).
7. Lau, K.W.; Chen, C.D.; Lee, H.L.; Izzul, A.A.; Asri-Isa, M.; Zulfadli, M.; Sofian-Azirun, M. Vertical distribution of *Aedes* mosquitoes in multiple storey buildings in Selangor and Kuala Lumpur, Malaysia. *Trop. Biomed.* **2013**, *30*, 36–45.
8. Sarip, A.G.; Foong, L. *Exploring the Perception of Lifestyle Housing Development in Malaysia*; The Asia Pacific Network for Housing Research: Gwangju, Korea, 2015.
9. New Straits Times. Malaysian House Buyers Increasingly Drawn to High-Rise Living. Available online: https://www.nst.com.my/news/2016/07/161815/malaysian-house-buyers-increasingly-drawn-high-rise-living (accessed on 15 April 2020).
10. Ministry of Health Malaysia. *Garis Panduan Halatuju Baharu Dalam Kawalan Denggi*; Vector Borne Disease Sector, Disease Control Division, Ministry of Health Malaysia: Putrajaya, Malaysia, 2014.
11. Muir, L.E.; Kay, B.H. *Aedes aegypti* survival and dispersal estimated by mark-release-recapture in northern Australia. *Am. J. Trop. Med. Hyg.* **1998**, *58*, 277–282.
12. Harrington, L.C.; Buonaccorsi, J.P.; Edman, J.D.; Costero, A.; Kittayapong, P.; Clark, G.G.; Scott, T.W. Analysis of survival of young and old *Aedes aegypti* (Diptera: Culicidae) from Puerto Rico and Thailand. *J. Med. Entomol.* **2001**, *38*, 537–547.
13. Verdonschot, P.F.M.; Besse-Lototskaya, A.A. Flight distance of mosquitoes (Culicidae): A metadata analysis to support the management of barrier zones around rewetted and newly constructed wetlands. *Limnologica* **2014**, *45*, 69–79.
14. Ministry of Health Malaysia. *Panduan Pengabutan (fogging) di Premis/ Bangunan Melebihi Lima (5) Tingkat*; Ministry of Health Malaysia: Putrajaya, Malaysia, 2010.
15. Jayathilake, H.; Wickramasinghe, M.; De Silva, N. Oviposition and vertical dispersal of *Aedes* mosquitoes in multiple storey buildings in Colombo district, Sri Lanka. *J. Vector Borne. Dis.* **2015**, *52*, 245–251.
16. Chadee, D.D. Observations on the seasonal prevalence and vertical distribution patterns of oviposition by *Aedes aegypti* (L.) (Diptera: Culicidae) in urban high-rise apartments in Trinidad, West Indies. *J. Vector Ecol.* **2004**, *29*, 323–330.
17. Liew, C.; Curtis, C.F. Horizontal and vertical dispersal of dengue vector mosquitoes, *Aedes aegypti* and *Aedes albopictus*, in Singapore. *Med. Vet. Entomol.* **2004**, *18*, 351–360.
18. Roslan, M.A.; Shafie, A.; Ngui, R.; Lim, Y.A.; Sulaiman, W.Y. Vertical infestation of the dengue vectors *Aedes aegypti* and *Aedes albopictus* in apartments in Kuala Lumpur, Malaysia. *J. Am. Mosq. Control Assoc.* **2013**, *29*, 328–336.
19. Wan-Norafikah, O.; Nazni, W.A.; Noramiza, S.; Shafa'ar-Ko'ohar, S.; Azirol-Hisham, A.; Nor-Hafizah, R.; Sumarni, M.G.; Mohd-Hasrul, H.; Sofian-Azirun, M.; Lee, H.L. Vertical dispersal of *Aedes* (Stegomyia) spp. in high-rise apartments in Putrajaya, Malaysia. *Trop. Biomed.* **2010**, *27*, 662–667.
20. Sairi, F.A.M.; Dom, N.C.; Camalxaman, S.N. Infestation profile of *Aedes* mosquitoes in multi-storey buildings in Selangor, Malaysia. *Procedia Soc. Behav. Sci.* **2016**, *222*, 283–289.
21. Lee, H.L. *Aedes* ovitrap and larval survey in several suburban communities in Selangor, Malaysia. *Mosq-Borne. Dis. Bull.* **1992**, *9*, 9–15.
22. Parija, S. Illustrated keys: Some mosquitoes of Peninsula Malaysia. *Trop. Parasitol.* **2014**, *4*, 70.
23. Pratt, H.D.; Stojanovich, C.J.; Magennis, N.J. *Workbook on Identification of* Aedes *Aegypti Larvae*; U.S. Department of Health, Education and Welfare: Atlanta, GA, USA, 1964.
24. IBM, C.R. *IBM SPSS Statistics for Windows, Version 25.0*; IBM Corp.: Armonk, NY, USA, 2017.

25. Mahmud, M.; Mutalip, H.; Lodz, N.A.; Shahar, H. Study on key *Aedes* spp breeding containers in dengue outbreak localities in Cheras district, Kuala Lumpur. *Int. J. Mosq. Res.* **2018**, *5*, 23–30.
26. Estrada-Franco, J.G.; Craig, G.B. *Biology, Disease Relationships, and Control of Aedes Albopictus*; Pan American Health Organization: Washington, DC, USA, 1995.
27. Dieng, H.; Saifur, R.G.M.; Hassan, A.A.; Salmah, M.R.C.; Boots, M.; Satho, T.; Jaal, Z.; Abu Bakar, S. Indoor-breeding of *Aedes albopictus* in northern peninsular Malaysia and its potential epidemiological implications. *PLoS ONE* **2010**, *5*, e11790.
28. Rozilawati, H.; Zairi, J.; Adanan, C.R. Seasonal abundance of *Aedes albopictus* in selected urban and suburban areas in Penang, Malaysia. *Trop. Biomed.* **2007**, *24*, 83–94.
29. Hashim, N.A.; Ahmad, A.H.; Talib, A.; Athaillah, F.; Krishnan, K.T. Co-breeding association of *Aedes albopictus* (Skuse) and *Aedes aegypti* (Linnaeus) (Diptera: Culicidae) in relation to location and container size. *Trop. Life. Sci. Res.* **2018**, *29*, 213–227.
30. Service, M.W. Mosquito (Diptera: Culicidae) dispersal—The long and short of it. *J. Med. Entomol.* **1997**, *34*, 579–588.
31. Nyamah, M.A.; Sulaiman, S.; Omar, B. Categorization of potential breeding sites of dengue vectors in Johor, Malaysia. *Trop. Biomed.* **2010**, *27*, 33–40.
32. Zainon, N.; Mohd-rahim, F.; Roslan, D.; Abd Samat, A.H. Prevention of *Aedes* breeding habitats for urban high-rise building in Malaysia. *Plan. Malays. J.* **2016**, *14*, 115–128.
33. Costero, A.; Edman, J.D.; Clark, G.G.; Scott, T.W. Life table study of *Aedes aegypti* (Diptera: Culicidae) in Puerto Rico fed only human blood versus blood plus sugar. *J. Med. Entomol.* **1998**, *35*, 809–813.
34. Dye, C. Vectorial capacity: Must we measure all its components? *Parasitol. Today* **1986**, *2*, 203–209.
35. Harrington, L.C.; Edman, J.D.; Scott, T.W. Why do female *Aedes aegypti* (Diptera: Culicidae) feed preferentially and frequently on human blood? *J. Med. Entomol.* **2001**, *38*, 411–422.
36. Hoffmann, E.J.; Miller, J.R. Reassessment of the role and utility of wind in suppression of mosquito (Diptera: Culicidae) host finding: Stimulus dilution supported over flight limitation. *J. Med. Entomol.* **2003**, *40*, 607–614.
37. Hales, S.; de Wet, N.; Maindonald, J.; Woodward, A. Potential effect of population and climate changes on global distribution of dengue fever: An empirical model. *Lancet* **2002**, *360*, 830–834.

Tropical Medicine and Infectious Disease

Article

The Emergence of Chikungunya ECSA Lineage in a Mayaro Endemic Region on the Southern Border of the Amazon Forest

Carla Julia da Silva Pessoa Vieira [1,†], David José Ferreira da Silva [1,†], Janaína Rigotti Kubiszeski [1], Laís Ceschini Machado [2], Lindomar José Pena [2], Roberta Vieira de Morais Bronzoni [1] and Gabriel da Luz Wallau [2,*]

1 Health Sciences Institute, Federal University of Mato Grosso, Sinop 78550-728, MT, Brazil; cjspvieira@gmail.com (C.J.d.S.P.V.); silvadjf@gmail.com (D.J.F.d.S.); janainarigottik@gmail.com (J.R.K.); robertabronzoni@gmail.com (R.V.d.M.B.)

2 Aggeu Magalhães Institute, Oswaldo Cruz Foundation, Recife 50670-420, PE, Brazil; laisceschini@gmail.com (L.C.M.); lindomar.pena@cpqam.fiocruz.br (L.J.P.)

* Correspondence: gabriel.wallau@cpqam.fiocruz.br or gabriel.wallau@gmail.com

† These authors contributed equally to this work.

Received: 19 May 2020; Accepted: 19 June 2020; Published: 26 June 2020

Abstract: Anthropic changes on the edges of the tropical forests may facilitate the emergence of new viruses from the sylvatic environment and the simultaneous circulation of sylvatic and urban viruses in the human population. In this study, we investigated the presence of arboviruses (arthropod-borne viruses) in the sera of 354 patients, sampled from February 2014 to October 2018 in Sinop city. We sequenced the complete genomes of one chikungunya virus (CHIKV)-positive and one out of the 33 Mayaro virus (MAYV)-positive samples. The CHIKV genome obtained here belongs to the East/Central/South African (ECSA) genotype and the MAYV genome belongs to the L genotype. These genomes clustered with other viral strains from different Brazilian states, but the CHIKV strain circulating in Sinop did not cluster with other genomes from the Mato Grosso state, suggesting that at least two independent introductions of this virus occurred in Mato Grosso. Interestingly, the arrival of CHIKV in Sinop seems to not have caused a surge in human cases in the following years, as observed in the rest of the state, suggesting that cross immunity from MAYV infection might be protecting the population from CHIKV infection. These findings reinforce the need for continued genomic surveillance in order to evaluate how simultaneously circulating alphaviruses infecting the human population will unfold.

Keywords: alphaviruses emergence; molecular epidemiology; human infection; arbovirus

---

1. Introduction

Arboviruses, arthropod-transmitted viruses, can successfully replicate in both invertebrates, vectors and vertebrate host cells [1]. These viruses are mostly transmitted by mosquitoes in sylvatic and urban environments, causing a large public health burden in several tropical countries around the globe [2]. WHO estimates that the dengue virus alone infects around 100–400 million people annually, mainly in the South American and Asian continents [3]. On the other hand, several human pathogenic arboviruses circulate in the sylvatic environment, with eventual spillover to the human population, such as Saint Louis encephalitis virus (SLEV) [4,5], West Nile virus (WNV) [6,7] and Mayaro virus (MAYV) [8,9].

Most known human-infecting arboviruses belong to the families Flaviviridae, Togaviridae and Peribunyaviridae [9,10]. Viruses from the Togaviridae family are zoonotic and epizootic pathogens that cause recurrent epidemics in both human and animal populations [11,12], such as the mosquito-borne alphaviruses chikungunya virus (CHIKV), O'nyong'nyong virus (ONNV), Ross River virus (RRV) and Mayaro virus (MAYV), and the viruses of the Western (WEEV), Eastern (EEEV) and Venezuelan (VEEV) equine encephalitides [12,13]. The arthritogenic alphaviruses MAYV and CHIKV were first isolated in the 1950s [14,15]. Since then, MAYV has been considered endemic and enzootic in South America and the Caribbean [16,17], while CHIKV emerged in Africa but spread globally, showing a much more complex evolutionary history. Phylogenomic analyses of CHIKV have revealed three distinct genotypes: West African (WA), East/Central/South African (ECSA) and Asian/Caribbean [18]. In Africa, both the WA and ECSA genotypes were circulating mainly enzootically among non-human primates, with occasional spillover to humans, while some ECSA lineages, such as the Indian Ocean lineage (IOL), expanded their range and have now been causing large outbreaks in the Asian continent. The Asian/Caribbean genotype is predominantly endemic/epidemic among human populations [19,20].

In Brazil, two CHIKV genotypes emerged in 2014: The Asian/Caribbean genotype in the Amapá state [21] and the ECSA genotype in the Bahia state [22]. Thereafter, a surge of human cases occurred; more than 500,000 cases have been reported by the Brazilian Ministry of Health [23]. Conversely, the neglected tropical Mayaro fever has no reports in the Brazilian epidemiological bulletins. However, studies have shown that 495 MAYV-infection cases have been detected in Brazil so far—more than half of all 901 cases reported in Latin America and the Caribbean [17]. These data show that the two viruses are constantly infecting the human population, with dramatically different dynamics but, until now, no studies have identified their co-circulation.

Chikungunya virus and MAYV circulate between an invertebrate vector and a vertebrate host [13], involving non-human primates and arboreal *Aedes* (for CHIKV) and *Haemagogus janthinomys* (for MAYV) mosquitoes [24,25]. Both the Asian/Caribbean and the ECSA lineages of CHIKV can circulate in urban cycles between humans and their main vectors *Aedes aegypti* and *Ae. albopictus*. Contrastingly, the WA lineage has never switched to an urban cycle and is maintained in a sylvatic cycle, including non-human primates and sylvatic *Aedes* mosquitoes, which are not anthropophilic but have caused sporadic human cases by biting humans in rural areas [13]. Similarly, MAYV causes disease after spillover transmission to nearby human populations inferred as incidental or dead-end hosts [26]. However, new evidence suggests that a sustained urban transmission is also possible since *Ae. aegypti* has been shown to transmit MAYV in laboratory conditions [27,28] and the MAYV genome has been detected in naturally infected *Ae. aegypti* specimens [29]. The high prevalence of *Ae. aegypti* in the urban environment in Brazil and other South American countries where Mayaro fever is endemic suggests that an urban cycle could be established [25]. Based on the vector species' habitat suitability and current knowledge about these viruses' distribution, it is possible that both alphaviruses spread and co-circulate, causing human outbreaks [30].

The whole-genome sequencing of clinical isolates during epidemics and interepidemic periods can be used to uncover the pathogen's natural history and spread pattern, providing background information to guide monitoring and prevention strategies, foster vaccine development and improve diagnostics and treatment [31]. The study of full-length arboviral genomic sequences provides access to evolutionary markers that allows more robust evolutionary inferences than viral fragments [32,33]. Herein, we report a genomic investigation and phylogenetic study based on the complete MAYV and CHIKV genomes sequenced from patients living on the southern border of the Brazilian Amazon rainforest.

2. Materials and Methods

2.1. Study Site

This study was carried out in Sinop city (11°50′53″ S and 55°38′57″ W), located 503 km north of the Mato Grosso state capital of Cuiabá, Midwest Brazil. The city was established in 1974, inside the Amazon rainforest, and today, it has an estimated population of 139,935 inhabitants. Sinop is situated in a transitional zone between the Amazon rainforest and Cerrado (Savannah) biomes. The region has a tropical climate, with a dry season of six months, and an average temperature of 26 °C and precipitation levels of over 2500 mm.

2.2. Sample Collection

Serum samples were collected during arboviral surveillance from February 2014 to October 2018. As inclusion criteria, we sought patients presenting clinical diagnoses of dengue, Zika or chikungunya infection, who reported arbovirus-like symptoms, such as fever, myalgia, headache and arthralgia. All of them were interviewed and signed an informed consent form.

2.3. Viral Detection—RT-PCR Assays

Viral RNA was extracted directly from the samples or cells' supernatant using a QIamp viral RNA mini kit (QIAGEN, Hilden, Germany), followed by a reverse transcription using 1 μL of random hexamers (Promega, Madison, MI, USA), 1 μL of Go Script Reverse Transcriptase (Promega) and 8 μL of RNA, for a 20 μL reaction, according to the manufacturers' instructions. A duplex-PCR was made using the genus-specific primers for alphaviruses (targeting nsP1 region) and flaviviruses (targeting NS5 region), followed by multiplex-semi-nested RT-PCR assays for species-specific identification [34], with a few modifications, as previously described [31].

2.4. Viral Detection—Isolation

The MAYV and CHIKV positive samples were inoculated into the cell culture of *Ae. albopictus* C6/36 and Vero cell lines; three passages were performed and then the supernatants were tested for viral nucleic acids by RT-PCR assays following [35].

2.5. Viral Genome Sequencing

Two PCR positive samples (one for CHIKV and one for MAYV) were assessed by real-time RT-PCR (rRT-PCR), employing the established protocols for CHIKV [36] and adding the MAYV primers used in the conventional PCR, in order to obtain Ct values to guide the genome sequencing strategy. Total RNA was extracted directly from the serum of the CHIKV positive sample and from the supernatant of the MAYV first passage in C6/36 cells, in order to perform the single-strand cDNA synthesis used for viral genomic amplification. RT-PCR reactions were carried out in a volume of 20 μL using Random Hexamers (Promega, Madison, WI, USA) and ProtoScript II Reverse Transcriptase (New England Biolabs, Ipswich, MA, USA), following the manufacturers' instructions.

Single strand cDNA from the CHIKV positive serum sample was submitted to whole genome amplification using multiplex-PCR [37], while single-strand cDNA from the MAYV sample was submitted to whole genome amplification by using multiplex RT-PCR, with primers designed using Primal Scheme [38]. Primer design was performed with two complete genomes of MAYV belonging to the L and N genotypes (GenBank accession number KY618127 and KP842812, respectively), both detected in South America (primers can be found in Supplementary Table S1). Sequencing libraries were prepared with the Nextera XT Library Prep Kit (Illumina, San Diego, CA, USA), using 2 ng of input cDNA derived from the CHIKV and MAYV multiplex PCRs, following the manufacturer's instructions. A MiSeq Reagent Kit V3 of 150 cycles (Illumina, San Diego, CA, USA) was used, employing a paired-end strategy, resulting in 75 bp reads separated by 350 bp. Samples were sequenced on an Illumina MiSeq platform at the Technological Platform Core at the Aggeu Magalhães Institute (IAM).

Trimmomatic v. 0.36 [39] was employed to remove low-quality reads, adapter sequences and primers. FastQC was used to assess the quality of the Illumina raw reads. These were subsequently mapped against the MAYV BeH473130/1988 (KY618133) and CHIKV BHI3734/H804698 (KP164568) reference genomes, using Bowtie 2 with the default parameters [40]. Consensus sequences of viral genomes were obtained through the Integrated Genome Viewer software [41] and the sequences were deposited in GenBank with the accession numbers: MH513597 and MT349960. The raw sequenced reads may be found in the European Nucleotide Archive under the project number PRJEB38124.

Single Nucleotide Polymorphisms (SNPs) calls were performed with samtools mpileup and vcf-annotate tools [42] with the following parameters: (vcf-annotate-filter Qual = 20/MinDP = 200/SnpGap = 20), including manual SNP checking.

2.6. Phylogenetic Analysis

Multiple sequence alignment was performed using MAFFT version 7 [43], through the MAFFT online server (http://mafft.cbrc.jp/alignment/software/); maximum-likelihood analysis (ML) was performed using PhyML version 3.0 [44] through the web site at http://www.atgc-montpellier.fr/phyml/. Coding regions corresponding to the complete genomes from Sinop were aligned with all published and available near-complete: CHIKV genomes (>8000 nucleotides), belonging to the WA, ECSA and Asian/Caribbean genotypes, totalizing 930 genomes; and MAYV genomes (>8000 nucleotides), belonging to the L, D and N genotypes, totalizing 66 genomes. We also performed phylogenetic analysis using only Brazilian genomes belonging to the CHIKV ECSA genotype, totalizing 79 genomes. All genomes and data were collected from ViPR [45] (http://www.viprbrc.org/), on 15 April 2020 (accession numbers of all genomes used in the analysis may be seen in Supplementary Table S2). The ML phylogenies were reconstructed by using the best-fit general time-reversible (GTR) model with invariant sites (+I) (GTR + I) for MAYV, and 4 gamma substitution rate categories (+G) (GTR + G + I) for CHIKV, all suggested as the most likely models to represent the data by the Smart Model Selection implemented in version 3.0 of PhyML online [46]. For the tree search operation, we used SPR, and statistical support for the phylogenetic nodes was evaluated by an SH-like approximate likelihood ratio test. Tree visualization and figure generation were performed with FigTree v1.4.4 [47].

2.7. Ethical Approval

This study was approved by the Research Ethics Committee of the Júlio Müller Hospital-Universidade Federal de Mato Grosso (UFMT) (approval number 288.172/2013) and the Research Ethics Committee of the UFMT (approval number 2.063.295/2018).

3. Results

Serum samples were collected from 354 patients exhibiting dengue-like illnesses, within 20 days of the onset of symptoms. Thirty-four patients were positive for alphaviruses—33 tested positive for MAYV (33/354; 9.3%) and one for CHIKV (1/354; 0.3%) (Figure 1). Seventy-eight samples were positive for flaviviruses and are described elsewhere (Supplementary Table S3).

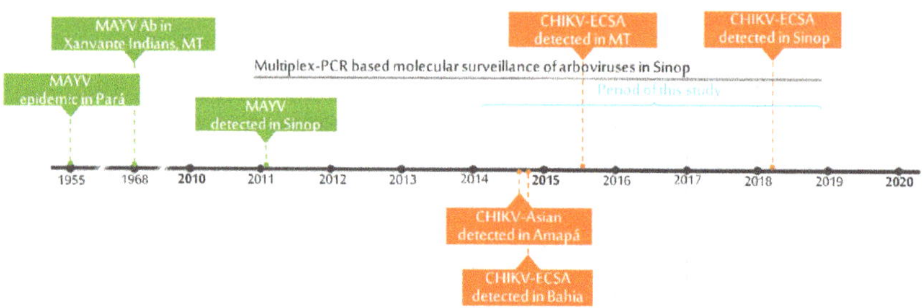

Figure 1. Timepoints of MAYV and CHIKV surveillance/detection in Sinop and nearby areas in the Amazon rainforest and the state of Mato Grosso.

The 33 Mayaro fever cases had a median age of 30 years; 18 (54.5%) were male and 15 (45.5%) were female. Clinical characteristics included mainly fever (n = 27), myalgia (n = 20), headache (n = 14) and arthralgia (n = 12). From the MAYV-positive samples, one was sequenced (BR/Sinop/H307/2015), which belonged to a 21-year-old female patient presenting with fever and myalgia, collected on the third day of symptoms, in March 2015. The other sample (BR/Sinop/H542/2018) was from a 20-year-old patient presenting with fever, myalgia, headache, arthralgia, retro-orbital pain and hematemesis, collected on the fourth day of symptoms, in February 2018 (Supplementary Table S4). Both patients were urban residents and reported no recent history of travel or access to urban parks, sylvatic or rural areas in temporal proximity to the emergence of the first symptoms.

The cycle thresholds (CTs) for the positive samples were 6.12 and 23.19 for the MAYV and CHIKV samples, respectively. We obtained 2,813,153 reads for the MAYV sample, in which 84.48% mapped to the reference genome, reaching an average coverage depth of 15,453.99; from the CHIKV positive sample, we obtained 1,123,424 reads, in which 96.47% mapped to the reference genome, reaching an average coverage depth of 7133.15 (Table 1).

Table 1. Epidemiology and sequencing data of two alphaviruses obtained from patients living in Sinop, MT.

Isolate ID	Age	Gender	Onset of Symptoms—dd/mm/yyyy	Collection Date—dd/mm/yyyy	Sample	qRT-PCR Ct [a]	Genbank Accession Code	Total No. Reads [b]	No. Mapped Reads (%) [b]	Depth of Coverage [b]	Genome Length (nt) [b]
H307 MAYV	21	F	14/03/2015	16/03/2015	C6/36 passage 1	6.12	MH513597	2,813,153	2,376,618 (84.48%)	15,453.99	11,147
H542 CHIKV	20	F	04/02/2018	07/02/2018	Serum	23.19	MT349960	1,123,424	1,083,817 (96.47%)	7133.15	11,492

Depth of coverage = number of mapped reads × 75 (read length)/reference genome length (11,534 for MAYV and 11,812 for CHIKV). [a] Ct, cycle threshold; qRT-PCR, quantitative reverse transcription PCR. [b] Genomic sequencing statistics (%) were calculated using KY618133 for MAYV (11,534 nt long) and KP164568 for CHIKV (11,812 nt long) as reference genome.

The maximum likelihood phylogenetic reconstruction of the obtained MAYV genome and a few other available MAYV genomes showed that the Sinop strain clustered within the L genotype and was closely related to the MAYV strains detected from mosquitoes in the 1960s in Pará state (PA) in the northern region of Brazil (Figure 2a,b), showing a high aLRT SH-like support of 1.0. Meanwhile, the CHIKV genome was placed in a polytomy within a previously described Brazilian subclade that is an offshoot of the East/Central/South African lineage [21], which also had a high aLRT SH-like support of 1.0 (Figure 3a). Interestingly, the CHIKV sequenced here did not cluster with other lineages from the state of Mato Grosso (Figure 3b).

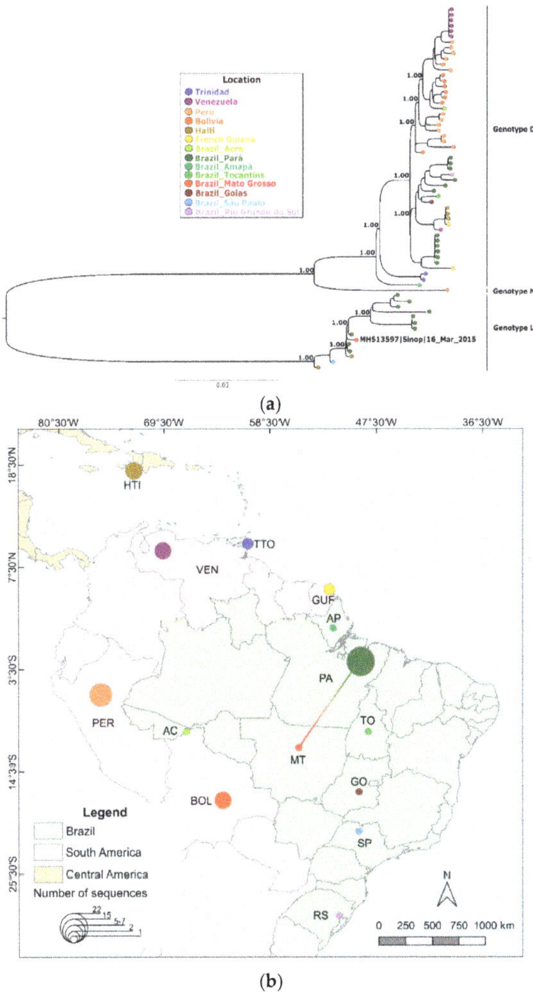

Figure 2. Full phylogenetic reconstruction of MAYV. (**a**) Phylogeny including all available MAYV genomes and positioning of MAYV genome obtained in this study using maximum likelihood of complete and draft genomes (>8 kb) available. (**b**) Distribution of the number of MAYV genomes sequenced per country and Brazilian state and colored edges showing a possible route of MAYV spread based on the most closely related MAYV genome from the phylogenetic analysis shown in panel A. Country and Brazilian state colors follow tip colors in panel A.

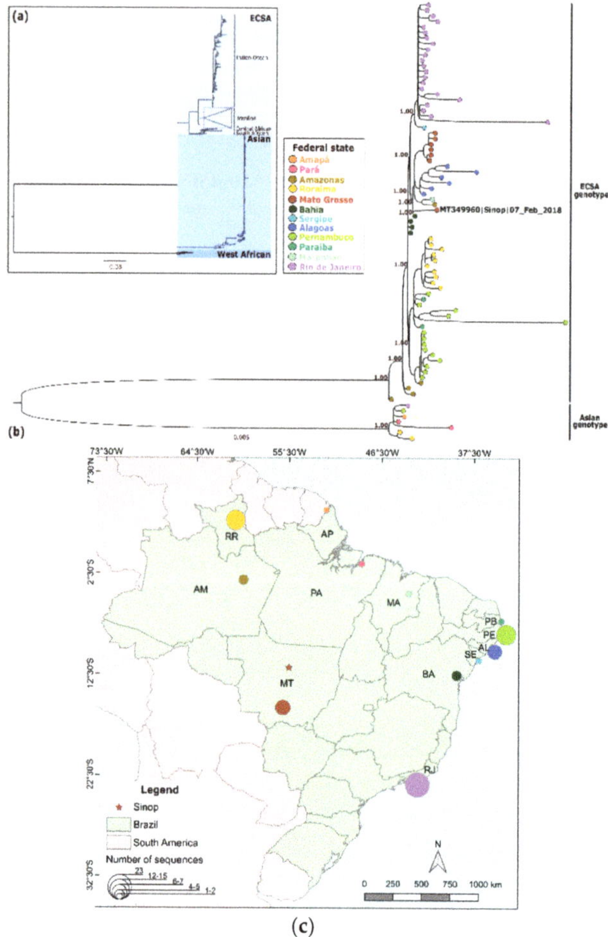

Figure 3. Phylogenetic reconstruction and positioning of CHIKV genome obtained in this study using maximum likelihood. (**a**) Full phylogeny including all available West African, East/Central/South African (ECSA)—Indian Ocean lineages and Asian/Caribbean using complete and draft genomes >8 kb. (**b**) Phylogenetic analysis focusing on Brazilian genomes of a subset of ECSA and Asian/Caribbean CHIKV genomic sequences from the full CHIKV tree. (**c**) Distribution of the number of CHIKV genomes sequenced per state. State colors follow tip colors in panel B.

We found a small number of SNPs per sample, varying from 17 in the CHIKV to 30 in the MAYV genome (Figure 3). For CHIKV, six SNPs found were non-synonymous and 11 were synonymous. For MAYV, seven SNPs found were non-synonymous and 22 were synonymous. Moreover, we found several specific mutations restricted to the genomes sequenced in this study: CHIKV presented four new amino acid mutations, two in nonstructural proteins (nsP2-T31I and nsP3-A388V) and two in an envelope protein (E3 T20I and H57R); MAYV presented two, both in the E1 protein (I425S and V427A) (Supplementary Table S5; Figure 4).

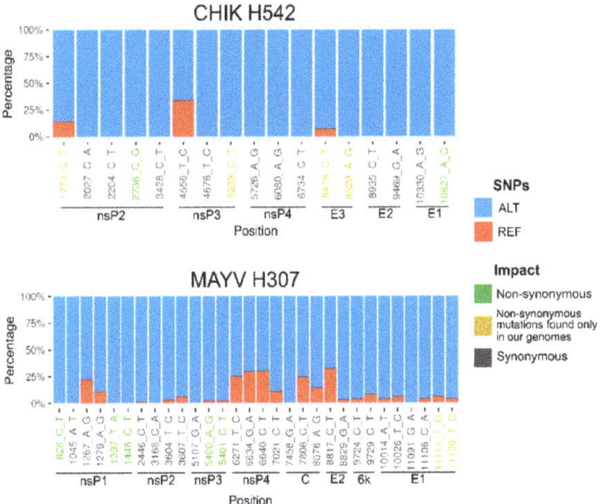

Figure 4. Single nucleotide polymorphism (SNP) analysis. Proportion of sequencing reads supporting the reference nucleotide (REF—red) and the SNPs (ALT—blue) from the CHIKV and MAYV genomes obtained in this study. Green, yellow and gray position names below each bar plot denote non-synonymous, non-synonymous mutations found only in our genomes and synonymous SNPs.

We detected two previously reported CHIKV amino acid mutations in structural proteins: E2-G60D, which contributes to the increased CHIKV fitness in *Ae. albopictus* and *Ae. aegypti* [48], and E1-T98A, which may increase the CHIKV infectivity for *Ae. albopictus* in the presence of E1-A226V substitution [49]. Five amino acid mutations, that were previously described to be under positive selection and to delineate the genotype L strains on vector infection, were revealed of MAYV. We also found nsP1-L518A, nsP3-A298P, nsP3-V386T, nsP4-A249K and E1-L300T [50].

4. Discussion

Arboviruses are widespread and diverse in Brazil and impose a large public health burden on the human population due to the abundance of sylvatic and urban vector species associated with poor sanitation and living conditions. As expected in such tropical environments, the simultaneous circulation and human co-infection of these viruses is common [51], imposing a great socioeconomic impact. Most people recover from arbovirus infection with no major sequelae, although life-threatening conditions can be identified such as hemorrhagic fever, Guillain–Barre syndrome and encephalitis [52]. Among a dozen arboviruses that currently circulate in Brazil and infect humans, dengue virus (DENV) has largely predominated, causing several thousand infections every year in the country. In 2019 alone, around 1.5 million cases were reported, and at least 782 people died from DENV infection [53]. Besides DENV, several other arboviruses have been causing human infection over the years, such as MAYV [8,9], Zika virus (ZIKV), yellow fever virus (YFV) and CHIKV, causing outbreaks with hundreds of thousands of people infected annually [54].

Most of the arboviral diagnoses in developing countries are based on clinical and epidemiological criteria, which leads to biased estimates of infection and produces underreports of less prevalent arboviruses that are not investigated in the molecular diagnostic routine. Such sub-notification can be clearly observed in Sinop from 2015 to 2019, where 7341 cases of DENV, 1324 of ZIKV and 15 of CHIKV were confirmed by laboratory tests, whereas the majority of the negative samples, around 44% of the cases investigated, were not tested for other arboviruses [55] (Supplementary Table S6). The confirmation of chikungunya diagnosis by Mato Grosso State Department of Health has been based on MAC-ELISA, while MAYV infection has not been investigated routinely yet. Meanwhile, molecular studies using human samples in Sinop have described the silent circulation of SLEV in 2011 and 2015 [56] and a MAYV annual incidence since 2011 [9]. Adding up to this scenario of simultaneous arbovirus circulation, here, we confirm the circulation of CHIKV and MAYV in human samples through molecular testing and genome sequencing.

Genome-wide phylogenetic analysis of MAYV confirmed the circulation of the L genotype in Sinop city, as previously demonstrated [9]. The obtained genome clustered with a genome from an isolate from the state of Pará (PA). Genotype D (widely dispersed) currently covers a broad geographic area from Trinidad and Tobago to Brazil, Peru, Bolivia, Venezuela, Haiti and French Guiana, with several available full-genome sequences (51 genomes). On the other hand, genotype L (limited) has only a limited number of complete genomes available, consisting of ten isolates obtained over the years 1955 to 1991 from PA and two recent isolates: one from Haiti and another from the State of São Paulo, imported from PA [57]. Due to the limited data, it is not possible to precisely determine the L genotype's dispersion throughout the Amazonian region and its borders. Other studies have also detected fragments of L and D genotypes in other municipalities in MT [58], whereas molecular studies in other border states with MAYV circulation have not been performed yet [59].

We detected two non-synonymous amino acid substitutions I425S and V427A in the MAYV E1 protein, but whether these substitutions played any role in the cell entry capacity and viral persistence is unknown. Alphavirus glycoproteins E1/E2 mediate host recognition and entry into the cell [60]. These proteins are useful in the development of vaccines and serodiagnostic assays as they hold most of the immunogenic epitopes [61,62]. However, the constant humoral immune pressure creates amino acid variations in these proteins that may lead to viral evasion, influence cross-neutralization activity and allow host-switching [63]. Reverse genetic studies are needed to determine whether any of these substitutions may cause major changes in viral fitness. Continued genome-based molecular investigation of MAYV is recommended in order to understand the spread and maintenance of MAYV in the Southern Amazon region and assess antigenic variations that might impact cross-immunity between alphaviruses and affect the sensitivity of diagnostic tests based on protein binding (immunoassays) [64].

Our CHIKV isolate belonged to the ECSA lineage, sharing 99.8% of its nucleotide identity with isolates from Bahia, a state in Northeast Brazil. The CHIKV genome from Sinop did not cluster with other genomes from MT [65]. The distinct evolutionary divergence from the monophyletic MT clade suggests that CHIKV was introduced into MT by at least two independent CHIKV ECSA variants, which are likely co-circulating in the state [66]. With the current data, the circulation of CHIKV genotypes other than ECSA in MT cannot be ruled out, especially in light of the circulation of the Asian/Caribbean genotype in the Amazon region (states of Pará, Roraima and Amapá), highlighting the need for continuous molecular surveillance in the region.

Genome analysis revealed mutations E1-T98A and E2-G60D in the glycoprotein of the CHIKV strain from Sinop. The E2-G60D amino acid change was reported to contribute to CHIKV's fitness increase in both *Ae. albopictus* and *Ae. aegypti* [48] and is currently detected in all the Asian/Caribbean and ECSA lineage genomes that are available. Meanwhile, the E1-T98A amino acid change has been found in all available ECSA genomes and may increase CHIKV infectivity for *Ae. albopictus* in the presence of E1-A226V substitution [49]. We did not detect other amino acid changes in the E1 glycoprotein previously associated with CHIKV fitness-enhancing in *Ae. aegypti* (E1-A226V, E1-K211E and E2-V264A) [67] and *Ae. albopictus* (E1-A226V, E2-I211T) [68,69], suggesting that the CHIKV lineage circulating in Sinop likely does not have a higher fitness compared to other CHIKV lineages circulating in Brazil. However, the ECSA genotype is currently spreading all over South America and infecting an abundant and diverse *Ae. aegypti* and *Ae. albopictus* population. Since the E1-T98A amino acid substitution is fixed in all ECSA strains from Brazil, including the one sequenced in this study, if the E1-A226V change occurs, CHIKV's lineage would likely gain increased fitness in *Ae. albopictus*. In this scenario, transmission of this virus could be exacerbated throughout Brazilian and South American territories, as *Ae. albopictus* is already spread over the entire continent [70]. Finally, other sample specific mutations were also detected in the Sinop CHIKV genome, but further genetic studies are required to understand their importance in the virus' fitness, if any.

As previously discussed, MAYV is endemic in the Southern Amazon region, where many outbreaks have been reported in the last decade [9,58,71]. Considering that alphavirus antibody cross-immunity has already been reported [72], people living in the Southern Amazon region may have alphavirus cross-immunity due to a previous infection by MAYV. In fact, it has been suggested that infection by CHIKV may confer cross-protection from MAYV, following the use of sera from CHIKV-exposed patients to cross-neutralize MAYV in vitro [73]. Such cross-immunity could be the reason why CHIKV only emerged in Sinop in 2018 and has not caused large outbreaks since then. Likewise, the restricted spread of CHIKV in other MAYV endemic areas of the Amazon has been pointed out [74]. However, a comprehensive serological investigation based on neutralization assays or new diagnostic tools must be performed in order to assess the role of alphavirus cross-immunity in such areas.

It is worth noting that, despite the underreports, the highest CHIKV incidence rate in the central-west region of Brazil has been reported in MT (387.6 cases per 100,000 inhabitants), also in 2018 [23]. Noticeably, reports in Sinop did not follow the same trend, both using PCR diagnostic and serological assays, as most of the cases concentrated in the southern cities of the state (Supplementary Table S6). Continuous molecular surveillance is required in order to monitor how the spread of recently emerged viruses will unfold [13].

5. Conclusions

Epidemiological surveillance based on genome-scale sequencing of the circulating viral strains is valuable for the prompt detection of adaptive mutations, which is essential for understanding transmission patterns, assessing the risk of emergence and intervening in vector control strategies [75]. In this study, we found new mutations in the strains of MAYV and CHIKV and observed that these strains clustered with genomes from geographically distant Brazilian states, suggesting that their spread occurred through infected patients that traveled between states. Further studies are encouraged in order to follow the spread of these viruses within the Southern Amazon region, in order to further understand the importance of mutations in the maintenance and spread of MAYV and CHIKV. This need is reinforced by the large outbreaks of CHIKV in Brazil and the underreporting of MAYV infections [76,77]. New genomic data can clarify the epidemiological characteristics, such as adaptation to vector spread and impact on human infection, where arboviruses of the same viral family co-circulate and may have cross antibody reactivity.

Supplementary Materials: The following are available online at http://www.mdpi.com/2414-6366/5/2/105/s1, Table S1. Primer sets used for whole MAYV genome amplification. Table S2a. Sixty-six complete Mayaro virus genomes, belonging to the D and L genotypes, used to reconstruct ML phylogenetic analysis. Table S2b. 930 complete and near-complete Chikungunya virus genomes, belonging to the West African, East/Central/South African and Asian/Caribbean genotypes, used to reconstruct ML phylogenies. Table S3. Chikungunya and Mayaro viruses distribution throughout the years. Table S4. Clinical and epidemiological features of patients who tested positive for MAYV and CHIKV in Sinop, MT, from 2014 to 2018. Table S5. SNPs called from each genome and associated statistics. Table S6. DENV, ZIKV and CHIKV notified and positive cases in Sinop, confirmed by the Mato Grosso State Department of Health, from 2014 to 2019.

Author Contributions: Conceptualization, R.V.d.M.B. and G.d.L.W.; Data curation, C.J.d.S.P.V. and D.J.F.d.S.; Formal analysis, C.J.d.S.P.V., D.J.F.S., and G.d.L.W.; Funding acquisition, L.J.P., R.V.d.M.B. and G.d.L.W; Investigation, C.J.d.S.P.V., D.J.F.d.S., J.R.K. and L.C.M.; Methodology, C.J.d.S.P.V., D.J.F.d.S., J.R.K., L.C.M., R.V.d.M.B., and G.d.L.W.; Project administration, R.V.d.M.B. and G.d.L.W.; Resources, L.J.P., R.V.d.M.B., and G.d.L.W.; Software, C.J.d.S.P.V. and G.d.L.W.; Supervision, R.V.d.M.B. and G.d.L.W.; Validation, C.J.d.S.P.V., D.J.F.d.S., J.R.K., and L.C.M.; Visualization, C.J.d.S.P.V. and G.d.L.W.; Writing—original draft, C.J.d.S.P.V., D.J.F.d.S., and G.d.L.W.; Writing—review & editing, C.J.d.S.P.V., D.J.F.d.S., J.R.K., L.C.M., L.J.P., R.V.d.M.B., and G.d.L.W. All authors have read and agreed to the published version of the manuscript.

Funding: This research was funded by Fundação de Amparo à Pesquisa do Estado de Mato Grosso (FAPEMAT/SES-MT/CNPq 002/2013, grant number 228398/2013 and FAPEMAT 007/2016, grant number 499331/2016), Fundação de Amparo à Ciência e Tecnologia do Estado de Pernambuco (FACEPE, grant number APQ-0078-2.02/16) and Conselho Nacional de Desenvolvimento Científico e Tecnológico-CNPq for the research grant number PQ-2 of Wallau, GL (303902/2019-1).

Conflicts of Interest: The authors declare no conflict of interest.

References

1. Ebel, G.D. Promiscuous viruses—how do viruses survive multiple unrelated hosts? *Curr. Opin. Virol.* **2017**, *23*, 125–129. [CrossRef]
2. Marcondes, C.B.; Contigiani, M.; Gleiser, R.M. Emergent and Reemergent Arboviruses in South America and the Caribbean: Why So Many and Why Now? *J. Med. Entomol.* **2017**, *54*, 509–532. [CrossRef]
3. World Health Organization Dengue and severe dengue. Available online: https://www.who.int/news-room/fact-sheets/detail/dengue-and-severe-dengue (accessed on 30 April 2020).
4. Rocco, I.M.; Santos, C.L.S.; Bisordi, I.; Petrella, S.M.C.N.; Pereira, L.E.; Souza, R.P.; Coimbra, T.L.M.; Bessa, T.A.F.; Oshiro, F.M.; Lima, L.B.Q.; et al. St. Louis encephalitis virus: First isolation from a human in São Paulo State, Brazil. *Rev. Inst. Med. Trop. Sao Paulo* **2005**, *47*, 281–285. [CrossRef]
5. Mondini, A.; Bronzoni, R.V.d.M.; Cardeal, I.L.S.; dos Santos, T.M.I.L.; Lázaro, E.; Nunes, S.H.P.; Silva, G.C.D.; Madrid, M.C.F.S.; Rahal, P.; Figueiredo, L.T.; et al. Simultaneous infection by DENV-3 and SLEV in Brazil. *J. Clin. Virol.* **2007**, *40*, 84–86. [CrossRef]
6. Vieira, M.A.C.e.S.; Aguiar, A.d.A.X.; De Borba, A.S.; Guimarães, H.C.L.; Eulálio, K.D.; de Albuquerque-Neto, L.L.; Salmito, M.d.A.; Lima, O.B. West Nile fever in Brazil: Sporadic case, silent endemic disease or epidemic in its initial stages? *Rev. Inst. Med. Trop. Sao Paulo* **2015**, *57*, 276. [CrossRef]
7. Martins, L.C.; Da Silva, E.V.P.; Casseb, L.M.N.; Da Silva, S.P.; Cruz, A.C.R.; De Sousa Pantoja, J.A.; De Almeida Medeiros, D.B.; Filho, A.J.M.; Da Cruz, E.D.R.M.; De Araújo, M.T.F.; et al. First isolation of west nile virus in brazil. *Mem. Inst. Oswaldo Cruz* **2019**, *114*. [CrossRef] [PubMed]
8. Terzian, A.C.B.; Auguste, A.J.; Vedovello, D.; Ferreira, M.U.; Da Silva-Nunes, M.; Sperança, M.A.; Suzuki, R.B.; Juncansen, C.; Jo, J.P.A.; Weaver, S.C.; et al. Isolation and characterization of Mayaro virus from a human in Acre, Brazil. *Am. J. Trop. Med. Hyg.* **2015**, *92*, 401–404. [CrossRef]
9. Vieira, C.J.d.S.P.; da Silva, D.J.F.; Barreto, E.S.; Siqueira, C.E.H.; Colombo, T.E.; Ozanic, K.; Schmidt, D.J.; Drumond, B.P.; Mondini, A.; Nogueira, M.L.; et al. Detection of Mayaro virus infections during a dengue outbreak in Mato Grosso, Brazil. *Acta Trop.* **2015**, *147*, 12–16. [CrossRef] [PubMed]
10. Ahmed, A.; Dietrich, I.; Desiree LaBeaud, A.; Lindsay, S.W.; Musa, A.; Weaver, S.C. Risks and challenges of arboviral diseases in Sudan: The urgent need for actions. *Viruses* **2020**, *12*, 81. [CrossRef] [PubMed]
11. Pfeffer, M.; Dobler, G. Emergence of zoonotic arboviruses by animal trade and migration. *Parasites and Vectors* **2010**, *3*, 35. [CrossRef]
12. Go, Y.Y.; Balasuriya, U.B.R.; Lee, C. Zoonotic encephalitides caused by arboviruses: Transmission and epidemiology of alphaviruses and flaviviruses. *Clin. Exp. Vaccine Res.* **2014**, *3*, 58. [CrossRef] [PubMed]
13. Levi, L.I.; Vignuzzi, M. Arthritogenic alphaviruses: A worldwide emerging threat? *Microorganisms* **2019**, *7*, 133. [CrossRef] [PubMed]
14. Casals, J.; Whitman, L. Mayaro virus: A new human disease agent. I. Relationship to other arbor viruses. *Am. J. Trop. Med. Hyg.* **1957**, *6*, 1004–1011. [CrossRef] [PubMed]
15. Mason, P.J.; Haddow, A.J. An epidemic of virus disease in Southern Province, Tanganyika Territory, in 1952–1953. *Trans. R. Soc. Trop. Med. Hyg.* **1957**, *51*, 238–240. [CrossRef]
16. Lorenz, C.; Freitas Ribeiro, A.; Chiaravalloti-Neto, F. Mayaro virus distribution in South America. *Acta Trop.* **2019**, *198*. [CrossRef]
17. Ganjian, N.; Riviere-Cinnamond, A. Mayaro virus in Latin America and the Caribbean. *Rev. Panam. Salud Pública* **2020**, *44*, 1. [CrossRef]
18. Chen, R.; Puri, V.; Fedorova, N.; Lin, D.; Hari, K.L.; Jain, R.; Rodas, J.D.; Das, S.R.; Shabman, R.S.; Weaver, S.C. Comprehensive Genome Scale Phylogenetic Study Provides New Insights on the Global Expansion of Chikungunya Virus. *J. Virol.* **2016**, *90*, 10600–10611. [CrossRef]
19. Weaver, S.C. Arrival of Chikungunya Virus in the New World: Prospects for Spread and Impact on Public Health. *PLoS Negl. Trop. Dis.* **2014**, *8*. [CrossRef]
20. Silva, J.V.J., Jr.; Ludwig-Begall, L.F.; Oliveira-Filho, E.F.d.; Oliveira, R.A.S.; Durães-Carvalho, R.; Lopes, T.R.R.; Silva, D.E.A.; Gil, L.H.V.G. A scoping review of Chikungunya virus infection: Epidemiology, clinical characteristics, viral co-circulation complications, and control. *Acta Trop.* **2018**, *188*, 213–224. [CrossRef]
21. Nunes, M.R.T.; Faria, N.R.; de Vasconcelos, J.M.; Golding, N.; Kraemer, M.U.G.; de Oliveira, L.F.; da Silva Azevedo, R.d.S.; da Silva, D.E.A.; da Silva, E.V.P.; da Silva, S.P.; et al. Emergence and potential for spread of Chikungunya virus in Brazil. *BMC Med.* **2015**, *13*, 102. [CrossRef]

22. Teixeira, M.G.; Andrade, A.M.S.; Da Costa, M.C.N.; Castro, J.S.M.; Oliveira, F.L.S.; Goes, C.S.B.; Maia, M.; Santana, E.B.; Nunes, B.T.D.; Vasconcelos, P.F.C. East/central/South African genotype chikungunya virus, Brazil, 2014. *Emerg. Infect. Dis.* **2015**, *21*, 906–908. [CrossRef] [PubMed]
23. Ministério da Saúde Ministério da Saúde, 2020. Monitoramento dos casos de arboviroses urbanas transmitidas pelo Aedes (dengue, chikungunya e Zika), Semanas Epidemiológicas 01 a 52. Available online: https://www.saude.gov.br/images/pdf/2020/marco/06/Boletim-epidemiologico-SVS-10.pdf (accessed on 8 May 2020).
24. Mayer, S.V.; Tesh, R.B.; Vasilakis, N. The emergence of arthropod-borne viral diseases: A global prospective on dengue, chikungunya and zika fevers. *Acta Trop.* **2017**, *166*, 155–163. [CrossRef] [PubMed]
25. Izurieta, R.O.; DeLacure, D.A.; Izurieta, A.; Hoare, I.A.; Reina Ortiz, M. Mayaro virus: The jungle flu. *Virus Adapt. Treat.* **2018**, *Volume 10*, 9–17. [CrossRef]
26. Gould, E.; Pettersson, J.; Higgs, S.; Charrel, R.; de Lamballerie, X. Emerging arboviruses: Why today? *One Heal.* **2017**, *4*, 1–13. [CrossRef]
27. Long, K.C.; Ziegler, S.A.; Thangamani, S.; Hausser, N.L.; Kochel, T.J.; Higgs, S.; Tesh, R.B. Experimental transmission of Mayaro virus by Aedes aegypti. *Am. J. Trop. Med. Hyg.* **2011**, *85*, 750–757. [CrossRef] [PubMed]
28. Wiggins, K.; Eastmond, B.; Alto, B.W. Transmission potential of Mayaro virus in Florida *Aedes aegypti* and *Aedes albopictus* mosquitoes. *Med. Vet. Entomol.* **2018**, *32*, 436–442. [CrossRef] [PubMed]
29. Pereira Serra, O.; Fernandes Cardoso, B.; Maria Ribeiro, A.L.; dos Santos, F.A.L.; Dezengrini Slhessarenko, R. Mayaro virus and dengue virus 1 and 4 natural infection in culicids from Cuiabá, state of Mato Grosso, Brazil. *Mem. Inst. Oswaldo Cruz* **2016**, *111*, 20–29. [CrossRef]
30. Lwande, O.W.; Obanda, V.; Lindström, A.; Ahlm, C.; Evander, M.; Näslund, J.; Bucht, G. Globe-Trotting *Aedes aegypti* and *Aedes albopictus*: Risk Factors for Arbovirus Pandemics. *Vector-Borne Zoonotic Dis.* **2020**, *20*, 71–81. [CrossRef]
31. Vieira, C.J.d.S.P.; Machado, L.C.; Pena, L.J.; de Morais Bronzoni, R.V.; Wallau, G.L. Spread of two Zika virus lineages in Midwest Brazil. *Infect. Genet. Evol.* **2019**, *75*, 103974. [CrossRef]
32. Forrester, N.L.; Palacios, G.; Tesh, R.B.; Savji, N.; Guzman, H.; Sherman, M.; Weaver, S.C.; Lipkin, W.I. Genome-Scale Phylogeny of the Alphavirus Genus Suggests a Marine Origin. *J. Virol.* **2012**, *86*, 2729–2738. [CrossRef]
33. Schneider, A.d.B.; Ochsenreiter, R.; Hostager, R.; Hofacker, I.L.; Janies, D.; Wolfinger, M.T. Updated phylogeny of Chikungunya virus suggests lineage-specific RNA architecture. *bioRxiv* **2019**, 698522. [CrossRef] [PubMed]
34. de Morais Bronzoni, R.V.; Baleotti, F.G.; Nogueira, M.R.R.; Nunes, M.; Figueiredo, L.T.M. Duplex Reverse Transcription-PCR Followed by Nested PCR Assays for Detection and Identification of Brazilian Alphaviruses and Flaviviruses. *J. Clin. Microbiol.* **2005**, *43*, 696–702. [CrossRef] [PubMed]
35. Guedes, D.R.; Paiva, M.H.; Donato, M.M.; Barbosa, P.P.; Krokovsky, L.; Rocha, S.W.S.; La Saraiva, K.; Crespo, M.M.; Rezende, T.M.; Wallau, G.L.; et al. Zika virus replication in the mosquito Culex quinquefasciatus in Brazil. *Emerg. Microbes Infect.* **2017**, *6*, e69. [CrossRef]
36. Lanciotti, R.S.; Kosoy, O.L.; Laven, J.J.; Panella, A.J.; Velez, J.O.; Lambert, A.J.; Campbell, G.L. Chikungunya virus in US travelers returning from India, 2006. *Emerg. Infect. Dis.* **2007**, *13*, 764–767. [CrossRef] [PubMed]
37. Machado, L.C.; de Morais-Sobral, M.C.; Campos, T.d.L.; Pereira, M.R.; de Albuquerque, M.d.F.P.M.; Gilbert, C.; Franca, R.F.O.; Wallau, G.L. Genome sequencing reveals coinfection by multiple chikungunya virus genotypes in a recent outbreak in Brazil. *PLoS Negl. Trop. Dis.* **2019**, *13*, e0007332. [CrossRef]
38. Quick, J.; Grubaugh, N.D.; Pullan, S.T.; Claro, I.M.; Smith, A.D.; Gangavarapu, K.; Oliveira, G.; Robles-Sikisaka, R.; Rogers, T.F.; Beutler, N.A.; et al. Multiplex PCR method for MinION and Illumina sequencing of Zika and other virus genomes directly from clinical samples. *Nat. Protoc.* **2017**, *12*, 1261–1266. [CrossRef]
39. Bolger, A.M.; Lohse, M.; Usadel, B. Trimmomatic: A flexible trimmer for Illumina sequence data. *Bioinformatics* **2014**, *30*, 2114–2120. [CrossRef]
40. Langmead, B.; Salzberg, S.L. Fast gapped-read alignment with Bowtie 2. *Nat. Methods* **2012**, *9*, 357–359. [CrossRef]
41. Robinson, J.T.; Thorvaldsdóttir, H.; Winckler, W.; Guttman, M.; Lander, E.S.; Getz, G.; Mesirov, J.P. Integrative genomics viewer. *Nat. Biotechnol.* **2011**, *29*, 24–26. [CrossRef]

42. Danecek, P.; Auton, A.; Abecasis, G.; Albers, C.A.; Banks, E.; DePristo, M.A.; Handsaker, R.E.; Lunter, G.; Marth, G.T.; Sherry, S.T.; et al. The variant call format and VCFtools. *Bioinformatics* **2011**, *27*, 2156–2158. [CrossRef]
43. Katoh, K.; Rozewicki, J.; Yamada, K.D. MAFFT online service: Multiple sequence alignment, interactive sequence choice and visualization. *Brief. Bioinform.* **2017**, bbx108. [CrossRef] [PubMed]
44. Guindon, S.; Dufayard, J.-F.; Lefort, V.; Anisimova, M.; Hordijk, W.; Gascuel, O. New Algorithms and Methods to Estimate Maximum-Likelihood Phylogenies: Assessing the Performance of PhyML 3.0. *Syst. Biol.* **2010**, *59*, 307–321. [CrossRef] [PubMed]
45. Pickett, B.E.; Sadat, E.L.; Zhang, Y.; Noronha, J.M.; Squires, R.B.; Hunt, V.; Liu, M.; Kumar, S.; Zaremba, S.; Gu, Z.; et al. ViPR: An open bioinformatics database and analysis resource for virology research. *Nucleic Acids Res.* **2012**, *40*, D593–D598. [CrossRef]
46. Lefort, V.; Longueville, J.-E.; Gascuel, O. SMS: Smart Model Selection in PhyML. *Mol. Biol. Evol.* **2017**, *34*, 2422–2424. [CrossRef] [PubMed]
47. Rambaut, A. *FigTree: Tree Figure Drawing Tool Version 1.4.3*; Institute of Evolutionary Biology, Ed.; University of Edinburgh: Edinburgh, UK, 2014.
48. Tsetsarkin, K.A.; McGee, C.E.; Volk, S.M.; Vanlandingham, D.L.; Weaver, S.C.; Higgs, S. Epistatic roles of E2 glycoprotein mutations in adaption of Chikungunya virus to Aedes albopictus and Ae. Aegypti mosquitoes. *PLoS ONE* **2009**, *4*, e90972. [CrossRef]
49. Tsetsarkin, K.A.; Chen, R.; Leal, G.; Forrester, N.; Higgs, S.; Huang, J.; Weaver, S.C. Chikungunya virus emergence is constrained in Asia by lineage-specific adaptive landscapes. *Proc. Natl. Acad. Sci. USA* **2011**, *108*, 7872–7877. [CrossRef]
50. Auguste, A.J.; Liria, J.; Forrester, N.L.; Giambalvo, D.; Moncada, M.; Long, K.C.; Morón, D.; de Manzione, N.; Tesh, R.B.; Halsey, E.S.; et al. Evolutionary and ecological characterization of mayaro virus strains isolated during an outbreak, Venezuela, 2010. *Emerg. Infect. Dis.* **2015**, *21*, 1742–1750. [CrossRef]
51. Pauvolid-Corrêa, A.; Soares Juliano, R.; Campos, Z.; Velez, J.; Nogueira, R.M.R.; Komar, N. Neutralising antibodies for mayaro virus in Pantanal, Brazil. *Mem. Inst. Oswaldo Cruz* **2015**, *110*, 125–133. [CrossRef]
52. Beckham, J.D.; Tyler, K.L. Arbovirus Infections. *Contin. Lifelong Learn. Neurol.* **2015**, *21*, 1599–1611. [CrossRef]
53. Brazilian Ministry of Health Boletins epidemiológicos. Available online: https://www.saude.gov.br/boletins-epidemiologicos (accessed on 8 May 2020).
54. Zanotto, P.M.d.A.; Leite, L.C.d.C. The Challenges Imposed by Dengue, Zika, and Chikungunya to Brazil. *Front. Immunol.* **2018**, *9*, 1964. [CrossRef]
55. Secretaria de Estado de Saúde de Mato Grosso Boletim Epidemiológico: Dengue, febre de chikungunya e febre pelo vírus Zika. Available online: http://www.saude.mt.gov.br/dengue/arquivos/526/documentos (accessed on 8 May 2020).
56. Moraes, M.M.; Kubiszeski, J.R.; Vieira, C.J.d.S.P.; Gusmao, A.F.; Pratis, T.S.; Colombo, T.E.; Thies, S.F.; Araujo, F.d.C.; Zanelli, C.F.; Milhim, B.H.G.d.A.; et al. Concomitant detection of Saint Louis encephalitis virus in two Brazilian States. *Mem. Inst. Oswaldo Cruz.* (under review).
57. Mota, M.T.O.; Vedovello, D.; Estofolete, C.; Malossi, C.D.; Araújo, J.P.; Nogueira, M.L. Complete genome sequence of mayaro virus imported from the Amazon Basin to São Paulo State, Brazil. *Genome Announc.* **2015**, *3*. [CrossRef] [PubMed]
58. Zuchi, N.; Da Silva Heinen, L.B.; Dos Santos, M.A.M.; Pereira, F.C.; Slhessarenko, R.D. Molecular detection of Mayaro virus during a dengue outbreak in the state of Mato Grosso, Central-West Brazil. *Mem. Inst. Oswaldo Cruz* **2014**, *109*, 820–823. [CrossRef]
59. Brunini, S.; França, D.D.S.; Silva, J.B.; Silva, L.N.; Silva, F.P.A.; Spadoni, M.; Rezza, G. High frequency of mayaro virus IgM among febrile patients, central Brazil. *Emerg. Infect. Dis.* **2017**, *23*, 1025–1026. [CrossRef]
60. Zhang, R.; Kim, A.S.; Fox, J.M.; Nair, S.; Basore, K.; Klimstra, W.B.; Rimkunas, R.; Fong, R.H.; Lin, H.; Poddar, S.; et al. Mxra8 is a receptor for multiple arthritogenic alphaviruses. *Nature* **2018**, *557*, 570–574. [CrossRef]
61. Chua, C.-L.; Sam, I.-C.; Chiam, C.-W.; Chan, Y.-F. The neutralizing role of IgM during early Chikungunya virus infection. *PLoS ONE* **2017**, *12*, e0171989. [CrossRef]
62. Fumagalli, M.J.; De Souza, W.M.; Romeiro, M.F.; de Souza Costa, M.C.; Slhessarenko, R.D.; Figueiredo, L.T.M. Development of an enzyme-linked immunosorbent assay to detect antibodies targeting recombinant envelope protein 2 of Mayaro virus. *J. Clin. Microbiol.* **2019**, *57*. [CrossRef] [PubMed]

63. Cook, J.D.; Lee, J.E. The Secret Life of Viral Entry Glycoproteins: Moonlighting in Immune Evasion. *PLoS Pathog.* **2013**, *9*. [CrossRef] [PubMed]
64. Powers, A.M.; Aguilar, P.V.; Chandler, L.J.; Brault, A.C.; Meakins, T.A.; Watts, D.; Russell, K.L.; Olson, J.; Vasconcelos, P.F.C.; Da Rosa, A.T.; et al. Genetic relationships among Mayaro and Una viruses suggest distinct patterns of transmission. *Am. J. Trop. Med. Hyg.* **2006**, *75*, 461–469. [CrossRef] [PubMed]
65. Santos, F.M.; Dias, R.S.; de Oliveira, M.D.; Costa, I.C.T.A.; Fernandes, L.d.S.; Pessoa, C.R.; da Matta, S.L.P.; Costa, V.V.; Souza, D.G.; da Silva, C.C.; et al. Animal model of arthritis and myositis induced by the Mayaro virus. *PLoS Negl. Trop. Dis.* **2019**, *13*, e0007375. [CrossRef]
66. Harsha, P.K.; Reddy, V.; Rao, D.; Pattabiraman, C.; Mani, R.S. Continual circulation of ECSA genotype and identification of a novel mutation I317V in the E1 gene of Chikungunya viral strains in southern India during 2015–2016. *J. Med. Virol.* **2020**, jmv.25662. [CrossRef] [PubMed]
67. Agarwal, A.; Sharma, A.K.; Sukumaran, D.; Parida, M.; Dash, P.K. Two novel epistatic mutations (E1:K211E and E2:V264A) in structural proteins of Chikungunya virus enhance fitness in Aedes aegypti. *Virology* **2016**, *497*, 59–68. [CrossRef] [PubMed]
68. Vazeille, M.; Moutailler, S.; Coudrier, D.; Rousseaux, C.; Khun, H.; Huerre, M.; Thiria, J.; Dehecq, J.-S.; Fontenille, D.; Schuffenecker, I.; et al. Two Chikungunya Isolates from the Outbreak of La Reunion (Indian Ocean) Exhibit Different Patterns of Infection in the Mosquito, Aedes albopictus. *PLoS ONE* **2007**, *2*, e1168. [CrossRef] [PubMed]
69. Tsetsarkin, K.A.; Vanlandingham, D.L.; McGee, C.E.; Higgs, S. A Single Mutation in Chikungunya Virus Affects Vector Specificity and Epidemic Potential. *PLoS Pathog.* **2007**, *3*, e201. [CrossRef] [PubMed]
70. Cunze, S.; Kochmann, J.; Koch, L.K.; Klimpel, S. Niche conservatism of Aedes albopictus and Aedes aegypti-Two mosquito species with different invasion histories. *Sci. Rep.* **2018**, *8*, 1–10. [CrossRef]
71. Mourão, M.P.G.; Bastos, M.D.S.; De Figueiredo, R.P.; Gimaque, J.B.L.; Dos Santos Galusso, E.; Kramer, V.M.; De Oliveira, C.M.C.; Naveca, F.G.; Figueiredo, L.T.M. Mayaro fever in the city of manaus, Brazil, 2007–2008. *Vector-Borne Zoonotic Dis.* **2012**, *12*, 42–46. [CrossRef]
72. Martins, K.A.; Gregory, M.K.; Valdez, S.M.; Sprague, T.R.; Encinales, L.; Pacheco, N.; Cure, C.; Porras-Ramirez, A.; Rico-Mendoza, A.; Chang, A.; et al. Neutralizing Antibodies from Convalescent Chikungunya Virus Patients Can Cross-Neutralize Mayaro and Una Viruses. *Am. J. Trop. Med. Hyg.* **2019**, *100*, 1541–1544. [CrossRef]
73. Webb, E.M.; Azar, S.R.; Haller, S.L.; Langsjoen, R.M.; Cuthbert, C.E.; Ramjag, A.T.; Luo, H.; Plante, K.; Wang, T.; Simmons, G.; et al. Effects of Chikungunya virus immunity on Mayaro virus disease and epidemic potential. *Sci. Rep.* **2019**, *9*, 1–12. [CrossRef]
74. Naveca, F.G.; Claro, I.; Giovanetti, M.; de Jesus, J.G.; Xavier, J.; Iani, F.C.d.M.; do Nascimento, V.A.; de Souza, V.C.; Silveira, P.P.; Lourenço, J.; et al. Genomic, epidemiological and digital surveillance of Chikungunya virus in the Brazilian Amazon. *PLoS Negl. Trop. Dis.* **2019**, *13*, e0007065. [CrossRef]
75. Fischer, C.; de Lamballerie, X.; Drexler, J.F. Enhanced Molecular Surveillance of Chikungunya Virus. *mSphere* **2019**, *4*, e00295-19. [CrossRef]
76. Figueiredo, L.T.M. Large outbreaks of chikungunya virus in Brazil reveal uncommon clinical features and fatalities. *Rev. Soc. Bras. Med. Trop.* **2017**, *50*, 583–584. [CrossRef] [PubMed]
77. De O'Mota, M.T.; Avilla, C.M.S.; Nogueira, M.L. Mayaro virus: A neglected threat could cause the next worldwide viral epidemic. *Future Virol.* **2019**, *14*, 375–377. [CrossRef]

Tropical Medicine and Infectious Disease

Article

Identification of Mosquito Bloodmeals Collected in Diverse Habitats in Malaysian Borneo Using COI Barcoding

Katherine I. Young [1,*], Joseph T. Medwid [1], Sasha R. Azar [2,3,4], Robert M. Huff [5], Hannah Drumm [6], Lark L. Coffey [6,7], R. Jason Pitts [5], Michaela Buenemann [8], Nikos Vasilakis [2,3,9,10,11], David Perera [12] and Kathryn A. Hanley [1]

1 Department of Biology, New Mexico State University, Las Cruces NM 88003, USA; jtmedwid@nmsu.edu (J.T.M.); khanley@nmsu.edu (K.A.H.)
2 Department of Pathology, University of Texas Medical Branch, Galveston, TX 77555, USA; srazar@utmb.edu (S.R.A.); nivasila@utmb.edu (N.V.)
3 Department of Microbiology and Immunology, University of Texas Medical Branch, Galveston, TX 77555, USA
4 Institute for Translational Sciences, University of Texas Medical Branch, Galveston, TX 77555, USA
5 Department of Biology, Baylor University, Waco, TX 76706, USA; Robert_Huff1@baylor.edu (R.M.H.); Jason_Pitts@baylor.edu (R.J.P.)
6 School of Veterinary Medicine, University of California Davis, Davis, CA 95616, USA; hdrumm@ucdavis.edu (H.D.); lcoffey@ucdavis.edu (L.L.C.)
7 Department of Pathology, Microbiology & Immunology, University of California Davis, Davis, CA 95616, USA
8 Department of Geography, New Mexico State University, Las Cruces, NM 88003, USA; elabuen@nmsu.edu
9 Center for Biodefense and Emerging Infectious Diseases, University of Texas Medical Branch, Galveston, TX 77555, USA
10 Center for Tropical Diseases, University of Texas Medical Branch, Galveston, TX 77555, USA
11 Institute for Human Infection and Immunity, University of Texas Medical Branch, Galveston, TX 77555, USA
12 Institute of Health and Communiti Medicine, Universiti of Malaysia Sarawak, Sarawak 94300, Malaysia; dperera@unimas.my
* Correspondence: kiy761@nmsu.edu

Received: 7 January 2020; Accepted: 24 March 2020; Published: 1 April 2020

Abstract: Land cover and land use change (LCLUC) acts as a catalyst for spillover of arthropod-borne pathogens into novel hosts by shifting host and vector diversity, abundance, and distribution, ultimately reshaping host–vector interactions. Identification of bloodmeals from wild-caught mosquitoes provides insight into host utilization of particular species in particular land cover types, and hence their potential role in pathogen maintenance and spillover. Here, we collected 134 blood-engorged mosquitoes comprising 10 taxa across 9 land cover types in Sarawak, Malaysian Borneo, a region experiencing intense LCLUC and concomitant spillover of arthropod-borne pathogens. Host sources of blood were successfully identified for 116 (87%) mosquitoes using cytochrome oxidase subunit I (COI) barcoding. A diverse range of hosts were identified, including reptiles, amphibians, birds, and mammals. Sixteen engorged *Aedes albopictus*, a major vector of dengue virus, were collected from seven land cover types and found to feed exclusively on humans (73%) and boar (27%). *Culex tritaeniohynchus* (n = 2), *Cx. gelidus* (n = 3), and *Cx. quiquefasciatus* (n = 3), vectors of Japanese encephalitis virus, fed on humans and pigs in the rural built-up land cover, creating potential transmission networks between these species. Our data support the use of COI barcoding to characterize mosquito–host networks in a biodiversity hotspot.

Keywords: mosquito; vector; host; bloodmeal; arbovirus; Borneo; land cover and land use change; *Aedes*; dengue virus

1. Introduction

Mosquito-borne viruses such as dengue virus (DENV), yellow fever virus (YFV), Zika virus (ZIKV), and chikungunya virus (CHIKV) have gained notoriety for causing explosive, global pandemics sustained in transmission between humans and the urban-living mosquitoes *Ae. aegypti* and *Ae. albopictus* [1–3]. However, maintenance of such human-endemic cycles are actually quite rare among mosquito-borne viruses, the majority of which are zoonotic [4–6]. In general, mosquito-borne viruses are maintained in one of three cycles defined by specific hosts, vectors, and land cover types: 1) enzootic cycles involve wildlife hosts and vectors in natural environments, 2) rural epizootic cycles involve domestic animals and vectors in agriculture or rangelands, and 3) urban cycles involve humans or urbanized wildlife as hosts and urbanized vectors found in cities or suburbs [4]. Although distinct, these transmission cycles can be linked by spillover, the transmission of an arbovirus from a reservoir host via a bridge vector to a novel host [1,6,7]. For example, DENV, CHIKV, YFV, and ZIKV all arose in ancestral, enzootic cycles and spilled over to establish human-endemic cycles. Spillover can sometimes require amplification, which is infection of transmission-competent hosts that are fed upon by vectors that bridge to an established cycle and to novel, dead-end hosts from which no further transmission occurs [4,5]. Japanese encephalitis virus (JEV) is the main cause of human viral encephalitis in Asia but is maintained in an enzootic transmission cycle between mosquitoes and wading birds. However, the *Culex* mosquitoes that act as vectors are generalists and their feeding habits enable transmission to pigs. In pigs, the virus can amplify to high titers, and subsequently transmit to other hosts, including humans who are dead-end hosts for this virus [8].

The ability of a given vector species to mediate spillover depends on the breadth of its host utilization and its distribution across different habitats. Mosquitoes that are host and habitat generalists are more likely to act as bridge vectors than are highly specialized mosquitoes [9–12]. A number of extrinsic and intrinsic factors influence the host preference of mosquitoes, including olfaction, physiology, genetics, climate, host body heat, host defensive behavior, and host density [12]. However, these traits are complex, and determining innate host preferences for mosquitoes experimentally via lab or field studies can be biased by the use of colonized mosquitoes, seasonal and circadian changes in host and vector density, and variation between mosquito populations, among other factors [9,12]. An alternative to the experimental approach is to collect blood-engorged mosquitoes from the wild and use molecular methods to identify host use. While bloodmeal identification may not provide a complete picture of host preference [9], it can provide an accurate representation of host utilization patterns within a specific location and timeframe.

The current study used bloodmeal analysis to characterize mosquito–host interactions in Malaysian Borneo, a biodiversity hotspot [13] that is experiencing extreme rates of land cover and land use change (LCLUC) [14–17]. Since 1980, there has been steady clearance of native forests in Borneo, including old-growth dipterocarp, peat swamp, heath, and mangrove forests, primarily for logging and conversion to large plantations [14–17]. Meanwhile, high-density urban areas have expanded and people in low-density rural regions continue to clear forested areas for subsistence farming [18]. LCLUC plays an important role in arbovirus transmission by reshaping vector–host interactions via effects on vector and host distributions, abundance, and diversity [6,19–22]. In Borneo, extensive spillover of arboviruses [23–25] and other mosquito-borne pathogens [26,27] has been documented in the past several decades. Of particular concern are a number of cases of spillover of sylvatic DENV into humans [28–31], as each spillover event has the potential to launch a new strain of this virus into human-endemic transmission.

Here, we opportunistically sampled bloodfed mosquitoes from mosquito collections in nine land cover types, ranging from primary forest to barren land, in Sarawak Malaysian Borneo in 2016 and 2017. From these samples we sought: 1) To identify mosquito bloodmeal sources, using cytochrome oxidase subunit I (COI) barcoding, and 2) to represent the detected host–vector networks across these nine land cover types, while acknowledging that, given the number of samples analyzed, this represents only a small subset of the complete network within each land cover.

2. Materials and Methods

2.1. Study Site and Design

The study represents a concatenation of three projects focused on mosquito collections within the Kuching, Samarahan, and Serian administrative divisions in Sarawak, Malaysian Borneo (central coordinates: 1.5533° N, 110.3592° E), between April 2016 and June 2017 (Figure 1). Sarawak is considered Tropical Rainforest (Köppen-Geiger classification [32]) and receives rainfall year-round. This region of Sarawak comprises a mosaic of land covers including developed cities, peri-urban to rural villages, expansive cash crop plantations, small subsistence farms, and primary and secondary forests.

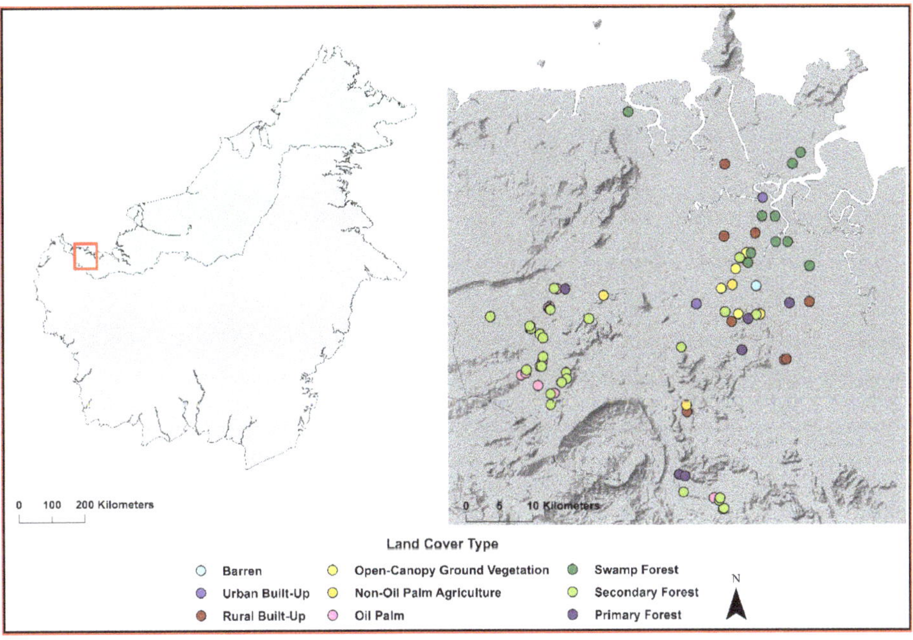

Figure 1. Study location in Sarawak, Malaysian Borneo is shown by the red box in the left panel. Sites where a bloodmeal was collected and the host was identified are shown in the right panel. The land cover type of each site sampled is designated by color.

The first project, termed the KK project, was a resampling of 15 sites across three land cover types in the village of Kampung Karu in Sarawak from April to May of 2016. Mosquito sampling was focused around homesteads, which are designated as rural built-up sites in this study, mixed agricultural fields, and secondary forests surrounding the village. The sampling method is thoroughly described by Young et al. (2017) [33]. Briefly, five sites in each land cover type were selected in the field based on observations made between 2013 and 2014 [33]. In 2016, these sites were resampled to measure changes in the abundance and diversity of mosquitoes and to evaluate land cover change between

2013/2014 and 2016 sampling years [33]. Thus, a total of 15 sites were sampled for this project, of which two yielded at least one blood-engorged mosquito for a total of two mosquitoes from this project.

The second project, called the OP project, began in June 2016 and ended in January 2017. Mosquito sampling was undertaken to assess the effects of oil palm insertion into forests on mosquito diversity, abundance, and distributions (Young et al. (2019) in revision for Ecosphere). Sampling was conducted in oil palm plantations, forests conterminous with those plantations, and the edge between these two land cover types. Google Earth satellite imagery from 25 July 2015 and 2 October 2015 was used to digitize >15 hectare oil palm plantations and adjacent forest boundaries in ArcGIS version 10.3.1 (ESRI, Redlands, California). Thirty-four sampling sites were randomly distributed at each of 10 isodistances, bands of equal distances measured from the edge of oil palm plantations and adjacent forests, within the boundaries of these land cover types; interior plantations to forest edges (−10 m, −50 m, −20 m, and −100 m), edge (0 m), and edge to the interior of contiguous forests (+10, +20, +50, +100, +500 m). A total of 340 sites were sampled for this project, of which 36 yielded at least one blood-engorged mosquito for a total of 43 mosquitoes from this project.

The third project, called the LC project, was undertaken from February to June 2017. Adult mosquito sampling was used to describe patterns in mosquito diversity, abundance, and distribution associated with different land cover types in Sarawak. Landsat 8 Operational Land Imager (OLI) imagery covering the study area from 13 October 2016, was downloaded via USGS EarthExplorer (Landsat 8 imagery courtesy of U.S. Geological Survey; https://earthexplorer.usgs.gov) and atmospherically and topographically corrected using ATCOR 3 [34,35](Atmospheric Corrections Software). Spectral and spatial features derived from the 8 multispectral OLI bands as well as topographic features derived from the Shuttle Radar Topography Mission (SRTM) Digital Elevation Model (DEM) [36,37] were combined in a 577-band layer stack. This layer stack was then classified using Random Forest and 600 Google Earth-based training/testing sites into eleven major land cover types: Urban built-up land, rural built-up land, agricultural land, primary forest, secondary forest, swamp forest, mangrove forest, open-canopy ground vegetation, barren land, and water. The overall classification accuracy was 80.2%. We subsequently generated 50 random points in each of eight of the land cover types; barren land, urban built-up land, rural built-up land, open-canopy ground vegetation, agricultural land, swamp forest, secondary forest, and primary forest. In total 132 sites were sampled from these eight land cover types, of which 36 yielded at least one blood-engorged mosquito for a total of 89 mosquitoes from this project.

At the completion of all projects, a total of nine land cover types were sampled; barren land, urban built-up land, rural built-up land, open-canopy ground vegetation, non-oil palm agricultural land, oil palm plantation, swamp forest, secondary forest, and primary forest; all are described in detail in Table S1 and Figure S1. Criteria for land cover designation, described in Table S1, was consistent across projects.

2.2. Mosquito Collection and Identification of Bloodfed Females

For the KK and OP projects, adult mosquitoes were collected using a backpack aspirator (Bioquip, Rancho Dominguez, CA, USA) [33], while for the LC project mosquitoes were collected at each site using a combination of gravid traps (Bioquip), CO_2 baited light traps (Clarke, St. Charles, IL, USA), BG sentinel traps (Bioquip), and aspiration via a Prokopac aspirator. Aspiration for the KK project was conducted for a duration of five minutes at nine trapping locations within each site and focused on open air, vegetation, and the body of the sampler [33]. For the OP and LC projects, aspiration was conducted around a 5 m radius from the center of each site by moving the aspirator in sweeping motions through the open air and directly onto or near ground vegetation and taller vegetation up to chest height, including trees and shrubs. For traps, one of each trap type was placed 5 m apart within the site and run until retrieval the following day. All BG sentinel traps were used with an octenol scent-bait (Bioquip). Gravid traps were placed over dark colored bins filled with hay infused water at a ratio of 3.8 g of hay: 1L water. Light traps were suspended at head height from either vegetation or existing structures and supplied with a yeast CO_2 bait made from 500 mL water, 7g INS yeast, and 50 g of coarse sugar, all acquired from a local grocery store [38].

Captured mosquitoes from all projects were transported in the collection containers used for each trap type. The LC project transported trap containers in a cooler with ice packs back to the Universiti of Malaysia Sarawak, where mosquitoes were killed by placing them at −20 °C, but the KK and OP projects transported trap containers directly without ice packs. Mosquitoes were then morphologically identified to the lowest taxonomic level possible using available keys [39–41] and the Walter Reed Biosystematics Unit system [42]. Female mosquitoes with visibly engorged abdomens were identified and their abdomens were separated and stored in individual tubes at −80°C for bloodmeal analysis. The LC project retained individual mosquito legs in a separate tube from all bloodfed females to be used for molecular identification of mosquitoes if needed; however, the KK and OP projects did not retain legs for this purpose.

2.3. Bloodmeal DNA Extraction and COI Amplification

DNA was extracted from the abdomen of blood-engorged mosquitoes according to a protocol described by Thiemann et al. (2012) using the Qiagen DNeasy Blood and Tissue Kit (Qiagen, Germantown, MD, USA) [43]. Twenty µl of proteinase K was added to each tube and a sterile pestle (Fisher Scientific, Hampton, NH) was used to gently crush the abdomen against the side releasing the bloodmeal. Next, 120 µl Buffer ATL was added and samples were incubated over night at 56 °C. Column extraction was then completed according to the manufacturer's instructions (Qiagen, Germantown, MD, USA).

Amplification of the cytochrome oxidase subunit I (COI) gene was conducted following one of two protocols. First, a nested PCR protocol described by Thiemann et al. (2012) was used on extracted DNA [43]. Briefly, the entire COI gene was amplified using primers that anneal to the tRNA regions flanking the gene at the 5′ and 3′ termini [43]. Four primer pairs were included in each reaction to cover a wide taxonomic breadth of hosts (Table S2) [43]. The PCR reaction was performed using Amplitaq (Applied Biosystems, Foster City, CA), and the volume of other reagents used are described in Table S2 under the tRNA protocol. Next an approximately 700 base pair fragment of the COI gene was amplified using HotStarTaq (Qiagen, Hilden, Germany) from each of the tRNA products using 6 internal primers: 3 forward and 3 reverse (Table S2) [43]. These forward and reverse primers were mixed in 1:1:2 ratios: VFmix is a 1:1:2 ratio of VF1, VF1d, VF1i, and VRmix is a 1:1:2 ratio of VR1, VR1d, VR1i (Table S3). All bloodmeal samples were screened using this protocol, and those that failed to generate products were then subject to a COI protocol described by Townzen et al. (2008) [44]. In brief, a 324 base pair region of COI was amplified from the extracted DNA using primers designed for a broad range of vertebrates and a recombinant Taq DNA polymerase (Invitrogen, Carlsbad, CA) (Table S2, Table S3). All PCR reactions were visualized on a 1.5% TAE agarose gel and included both negative and positive controls; the latter included DNA extracted from the blood of one amphibian (*Anaxyrus boreas*) (gift of Dr. Jamie Voyles, University of Nevada Reno IACUC 00698), one bird (*Melopsittacus undulates)* (gift of Dr. Timothy Wright, New Mexico State University (NMSU) IACUC 2016-006), one mammal (*Ovies aries*) (gift of Dr. Ryan Ashley, NMSU IACUC 2017-020), and one reptile (*Acrantophis spp.*) (gift of the Alamagordo Zoo, Alamagordo, New Mexico) species.

2.4. Sequence Analysis and Host Identification

Sanger sequencing was performed on samples with PCR products of the expected amplicon size. For samples amplified using the Thiemann protocol [43], the VFli primer was used for sequencing. Samples amplified using the Townzen protocol [44] were sequenced using both forward and reverse primers. Returned sequences were trimmed and, if amplified using the Townzen protocol, forward and reverse reads were aligned using ClustalW in Geneious software (Newark, NJ). Alignments were used to generate a consensus sequences for these samples. All samples were BLAST-searched against BOLD and NCBI databases. Generally, sequences with less than 3% sequence divergence are considered conspecific in DNA barcoding [45,46]. However, this designation is arbitrary, and it can be appropriate to use more broad sequence divergence criteria [47]. For host identification, sequences with ≥95% sequence identity were considered conspecific so long as the host identified was an endemic species of Borneo and frequently found in the land cover detected. Sequences with ≥80% identity were considered congeneric and between 77%–80% identity confamilic. However, if sequences had ≥80% identity to a species that is not known to occur in Borneo, or the surrounding region, they were considered confamilic.

2.5. Molecular Confirmation of Bloodfed Mosquito Species

To confirm morphological identifications of bloodfed mosquitoes, we attempted COI barcoding using a protocol described by Young et al. (2017) [33] from the LC project leg tissues. Briefly, total DNA and RNA was extracted using Trizol (Invitrogen, Pittsburgh, PA). Primers described by Folmer et al. (1998) [48] were then used in a PCR reaction to amplify a 710 base pair region of the COI gene. If the PCR reaction failed to yield a product, RNA was used in an RT-PCR reaction with the same primers to create cDNA for this region of the COI gene. The resulting products were then sequenced using both the forward and reverse primers, consensus sequences were generated, and resulting sequences were BLAST-searched against the BOLD and NCBI databases (Young et al. 2017) (Supplementary file SF1).

2.6. COI Cloning and Molecular Identification of Mosquitos from Bloodmeal DNA

Using the methods described in Section 2.5 in combination with morphological identification, three mosquitoes in which a bloodmeal had been amplified could be confidently identified to the level of species, two mosquitoes to genus, and one to family. To attempt to achieve more species-level identifications, we subjected a subset of 20 samples to cloning prior to sequencing. First, a PCR reaction was performed using primers that amplify an approximately 250 bp region of mosquito COI (Table S2) and 2x GoTaq Green PCR Master mix (Promega, Madison, WI) (Table S4). After PCR amplification, cloning using 2 µl of each reaction product was performed with the TOPO TA cloning kit (ThermoFisher Scientific, Waltham, MA) according to the manufacturer's protocol. Colonies were picked and placed into 3 mL of LB+Amp (100 ug/mL) liquid media and incubated at 37 °C for 16 h with shaking at 220 rpm. Miniprep liquid cultures (GeneJET, ThermoFisher Scientific, Waltham, MA) were created, diluted to 100 ng/µl, and then sequences at the UT Austin Sequencing Facility. These sequences were next compared to the BOLD and NCBI databases to confirm mosquito species. Because this amplicon was shorter than the traditional, 500-bp barcodes generated above, many of these sequences provided lower-level resolution. If a sample had been identified to species a priori using morphology and the sequence of the cloned amplicon had ≥98% identity to that species, we considered our morphological ID to be confirmed. If we had been unable to pinpoint a species-level ID and the sequence had equal percent identity across multiple species within the same genus, we considered this genus the lowest level of taxonomy achieved. If our sequence had equal percent identity across multiple genera, it was considered unresolved and the morphological identification was used as the final identification.

2.7. Descriptions of Host and Vector Networks

Due to the small number of bloodfed mosquitoes collected, we did not attempt statistical analyses of these data. To visualize host and vector networks, associations between vectors and hosts detected from bloodmeal identification, alluvial plots were created for each land cover type from which >5 bloodmeals were identified. Alluvial plots show the frequency of connections between categorical groups. The nodes of the plot, i.e., the black bars on the left and right sides, represent a category of interest, mosquito and host taxa in this case. The size of the node is relative to the proportion that each category represents of total samples. The flows of the plot, colored fans in the center of each plot, represent the connections between nodes and the size of each flow is proportional to the frequency at which those connections are made. Alluvial plots were created using RAWGraphics [49].

3. Results

3.1. Identification of Mosquito Bloodmeal Sources

Of 7299 mosquitoes collected from all projects, 134 were blood-engorged females, representing <2% of the total mosquitoes sampled. The relatively sparse yield of bloodfed mosquitoes is typical of studies in natural, tropical habitats [50–55]. For example, Brown et al. collected 127 bloodfed mosquitoes using three trap types in Sabah, Malaysian Borneo and were able to identify host species in 30% of these [52]. The small total number of blood fed mosquitoes in this study and the variation in sample size across land cover types, ranging from 1 mosquito in barren land to 42 mosquitoes in secondary forests, precluded statistical analysis of arbovirus spillover risk to humans.

The 134 specimens analyzed included eight genera of mosquitoes: *Aedes, Anopheles, Armigeres, Culex, Mansonia, Topomyia, Tripteroides, Uranotaenia*; one group identified to the Aedini tribe; and one specimen only identified to the Culicidae family. Engorged mosquitoes were collected from 74 sites in 9 land cover types in Sarawak. Four of these land cover types, barren, urban, open canopy ground vegetation and swamp forest, were sampled only in the LC project. Oil palm agriculture was sampled only in the OP project. Rural built-up (LC = 8 sites, KK = 1 site), non-oil palm agriculture (LC = 3 sites, KK = 1 site), secondary forest (OP = 24 sites, LC = 5 sites), and primary forests (LC = 3 sites, OP = 3 site) were sampled in two separate projects. However, since land cover classifications were consistent across projects, this is unlikely to confound consideration of networks within a land cover. Of the sampling methods used, 76% of the bloodfed mosquitoes were captured via aspiration, 19% via gravid traps, 4% via BG Sentinel traps, and 1% via light traps. Of the 134 blood fed mosquitoes, hosts were identified from 87% (n = 116).

Twenty-one unique host families were detected, including amphibians: Dicroglossidae Megophryidae, Ranidae; birds: Columbidae, Fringillidae, Phasianidae, Pittidae, Remizidae, Strigidae, Timaliidae; mammals: Canidae, Cervidae, Cynocephalidae, Hominidae, Muridae, Suidae, Tarsiidae, Tragulidae, and reptiles: Agamidae, Scincidae, Varanidae. Host identity was resolved to species for 92 specimens, to genus for 15, and to family for nine. No evidence of multiple hosts was detected in any mosquito.

With 62 specimens, *Culex* mosquitoes made up the majority of the collection and also encompassed the widest range of hosts including, reptilian (15%), amphibian (1%), avian (32%), and mammalian (52%) taxa. Similarly, a wide range of hosts were detected from 16 *Uranotaenia* specimens, with over half coming from reptilian and amphibian hosts (81%), 13% from avian hosts, and 6% from mammalian hosts. Humans were the only hosts detected from 16 mosquitoes identified only to the Aedini tribe. All sixteen *Aedes* samples collected were identified as *Ae. albopictus*. Seventy-five percent of *Ae. albopictus* bloodmeals identified were humans, and the remaining from *Sus scrofa* (wild boar) and *Sus barbatus* (bearded pig). Only a single sample each was collected from the remaining mosquito genera and reptiles and mammals were detected as hosts for these (Table 1, Supplementary file SF1).

Table 1. Description of hosts detected via cytochrome oxidase subunit I (COI) bloodmeal barcoding from specified mosquito genera or tribes in Sarawak. Whole numbers represent counts of host species detected. Percentages represent the proportion of bloodmeals detected from specified taxonomic groups for different mosquito genera.

Host	Common Name	*Aedes*	*Anopheles*	*Armigeres*	*Culex*	Culicidae	*Mansonia*	*Topomyia*	*Tripteroides*	*Uranotaenia*	Aedini
Amphibian					1%					63%	
Hylarana spp.	Golden-backed frog				1						
Leptobrachium hendricksoni	Spotted litter frog									3	
Limnonectes spp.	Forked-tongued frog									6	
Rana spp.	Brown frog									1	
Avian					32%					13%	
Ketupa ketupu	Buffy fish owl				1						
Fringillidae	Finches				1						
Gallus gallus	Red jungle fowl				12						
Macronous spp.	Pin-stripedtit-babbler				1						
Pitta nympha	Fairy pitta				4						
Remiz spp.	Eurasian pendulines									2	
Streptopelia chinensis	Spotted dove				1						
Mammal		100%	100%	100%	52%	100%	100%			5%	100%
Canis lupus familiaris	Domestic dog				2						
Cephalopachus bancanus	Horsfields tarsier				1						
Cynocephalus variegatus	Malayan colugo				1	1					
Homo sapiens	Human	12			10		1				16
Muntiacus muntjak	Barking deer			1							
Rattus tiomanicus	Malaysian field rat				1						
Sus barbatus	Bornean bearded pig	1			5					1	
Sus scrofa	Wild boar	3			12						
Tragulus spp	Mouse deer		1								
Reptile					15%			100%	100%	19%	
Agamidae	Dragon lizard				4			1			
Draco spp.	Flying lizard				2				1		
Scincidae	Skink									3	
Varanus salvator	Water monitor				3						
TOTALS		16	1	1	62	1	1	1	1	16	16

3.2. Identification of Mosquitoes Collected

Molecular taxonomy using COI barcoding was attempted on 104 of the bloodfed mosquito samples from which hosts had been identified. Eighty-six of these samples were attempted using DNA extracted from mosquito legs using the protocol described in Section 2.5, and 18 samples were attempted using DNA extracted from the mosquito abdomen using the protocol described in Section 2.6. The remaining samples were not subjected to COI barcoding because they were either easily identified morphologically or were morphologically identified similarly to the other submitted samples. Sequences were generated for 24 samples with lengths ranging from 267 to 619 bp; however, only five mosquito specimens could be identified to species with these sequences, including *Cx. tritaeniorhynchus* (N = 2), *Cx. gelidus* (N = 2), and *Ae. albopictus* (N = 1). Four samples were identified to genus, *Culex* (N = 1) and *Anopheles* (N = 1), tribe, Aedini (N = 1), or family, Culicidae (N = 1) (Supplementary file SF1). The remaining sequences had BLAST matches to multiple genera with ≥90% sequence identity, indicating an unresolved molecular identification.

3.3. Mosquito–Host Networks by Land Cover Type

Alluvial plots were created from those land covers with >5 mosquito–host interactions detected to visualize the relative frequency of connections between the sampled mosquitoes and hosts in the land cover types sampled. Barren, urban built-up, and rural built-up comprised three of the six human-modified land cover types in the study. Only one engorged mosquito, *Ae. albopictus*, was collected in barren land, and this mosquito had fed on a human. Four engorged *Culex quinquefasciatus* were collected in urban built-up land, three of which had fed on red jungle fowl and one on spotted dove. Within the rural built-up land cover type, seven host taxa were detected from 12 *Cx. quinquefasciatus*, six *Cx. gelidus*, two *Cx. tritaeniorhynchus*, 12 unidentified *Culex* and three *Ae. albopictus*. *Culex* species fed on mammals (70%; wild boar, humans, bearded pigs, dogs, and a Malaysian field rat), birds (27%, all red jungle fowl), and amphibians (3%, a golden-backed frog). Three *Ae. albopictus* in this land cover type fed on wild boar (Figure 2A).

Open-canopy ground vegetation, non-oil palm agriculture, and oil palm agriculture comprised the remaining three human-modified land cover types in the study. Three mammals were detected as hosts from six mosquitoes in the open-canopy ground vegetation land cover type. Bearded pig was identified as a host for one *Ae. albopictus* and one *Uranotaenia* sp., wild boar was identified as a host for two *Culex* sp., and humans were detected as a host from two *Ae. albopictus* (Figure 2B). Land under several agricultural crops, including cocoa (n = 1), durian (n = 1), fish (n = 1), and lime (n = 1) farms, made up the non-oil palm agriculture land cover type. In these agricultural lands, humans were identified from four *Culex* specimens and one *Ae. albopictus* (Figure 2C). Additionally, a buffy fish owl was identified as the host for one *Culex* sp. (Figure 2C). In oil palm plantations, humans were detected as hosts from five *Ae. albopictus* specimens and a forked-tongued frog from one *Uranotaenia* sp. (Figure 2D).

Three categories of forest were sampled in this study. In swamp forests, the most common mosquito collected was morphologically identified as *Verrallina*, but as molecular analyses to date have been inconclusive, these are here listed simply as Aedini. Humans were the only hosts detected from these 16 specimens (Figure 2E). Three *Culex* sp. from swamp forests fed on humans (33%) and water monitor lizards (67%) (Figure 2E). Amphibians were the only hosts detected from the *Uranotaenia* mosquitoes in this land cover type, including one spotted-litter frog and one brown frog (Figure 2E).

The majority of engorged mosquitoes were collected in secondary forests, while only 5 bloodmeals were identified from primary forest, thus these land cover types were combined into a single alluvial plot (Figure 2F). *Uranotaenia* was the most common genus from secondary forests; hosts for this genus of mosquitoes included reptiles (27%, skink), amphibians (64%, spotted litter frog and forked-tongued frog), and birds (9%, Eurasian penduline tit) (Figure 2F). *Culex* in secondary forests fed on birds (42%, pin-striped tit-babbler and fairy pitta), reptiles (50%, agamid lizard, flying lizard, and a water monitor), and mammals (8%, Malaysian colugo) (Figure 2F). Humans were detected as hosts for two *Ae. albopictus* specimens in secondary forests (Figure 2F). The remaining mosquito genera were collected as singleton samples and hosts encompassed a broad array of taxa (Figure 2F).

Primary forests yielded the only non-human primate bloodmeal, from a Horsfield's tarsier, detected in this study. This bloodmeal was isolated from a single *Culex* specimen (Figure 2F). Unfortunately, COI barcoding results have not enabled a species-level identification of this specimen. *Culex* sp. also fed upon a finch and an agamid lizard in this land cover type (Figure 2F). A single *Ae. albopictus* collected in primary forests fed on human, and a single *Uranotaenia* mosquito fed upon a Eurasian penduline tit (Figure 2F).

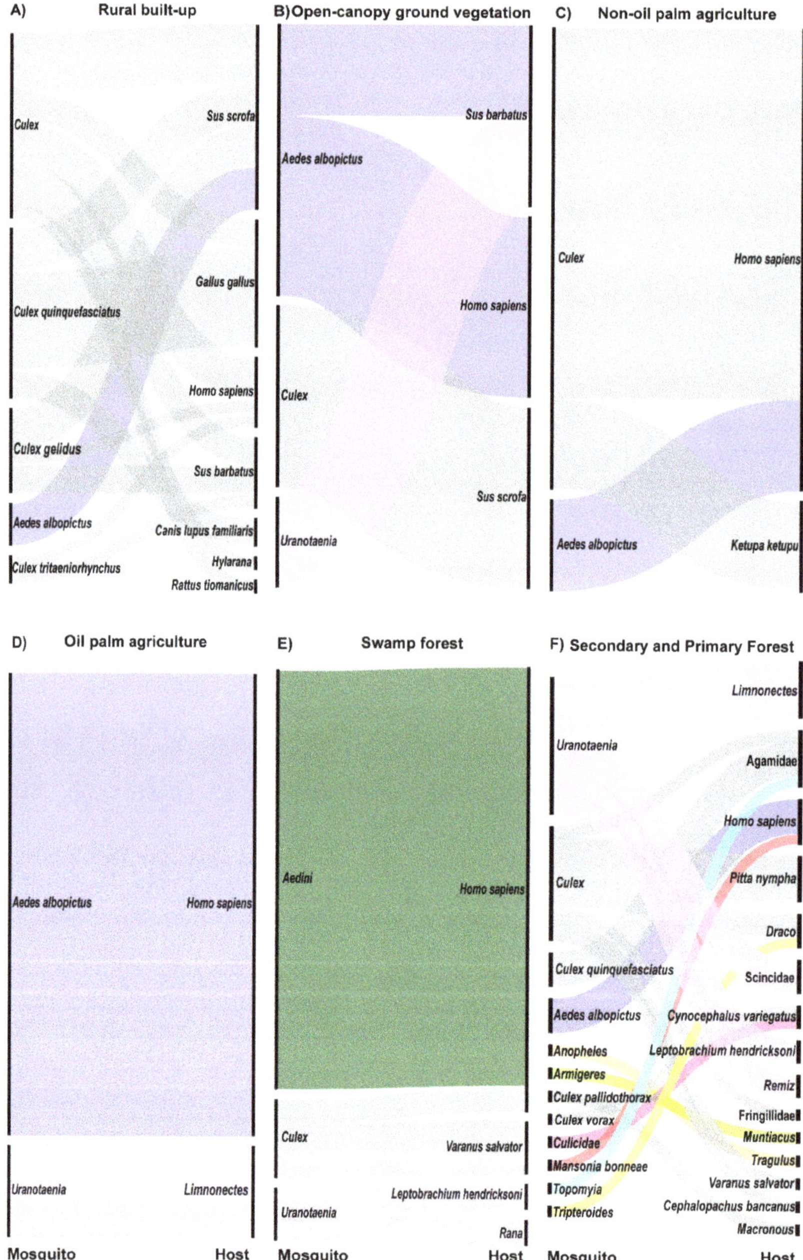

Figure 2. Alluvial plot showing mosquito and host networks at the land cover level for all land covers in which >5 bloodfed mosquitoes were collected (**A**) rural built-up (n = 36 mosquitoes), (**B**) open-canopy ground vegetation (n = 6), (**C**) non-oil palm agriculture (n = 6), (**D**) oil palm agriculture (n = 6), (**E**) swamp forests (n = 21), (**F**) secondary and primary forests (n = 36). Left nodes represent mosquito and right nodes represent host taxonomic groups. Colors of alluvial fans are consistent across figures and indicate mosquito genus.

3.4. Site Level Networks Between Mosquitoes and Hosts

For spillover of arboviruses to occur, vectors and hosts must be in close proximity. This may translate to a fine spatial scale, such as the site level, given the limited dispersal distances of most mosquitoes [56]. Therefore, sites were identified where either 1) a single mosquito taxonomic group was feeding on multiple host taxa or 2) two mosquito taxonomic groups were feeding on a single host taxon.

The first scenario, in which a single mosquito group fed on multiple taxa, played out at sites within the urban built-up (N = 1), rural built-up (N = 2), non-oil palm agriculture (N = 1), and secondary forest (N = 2) land cover types (Table 2). At site U_009, *Culex quinquefasciatus* fed on two avian species, red jungle fowl and spotted dove. Rural site R_015 had a network between four different *Culex* taxa and multiple hosts (Figure 3). One *Cx. gelidus* fed on humans, two on bearded pigs, and three on wild boar. Two *Cx. tritaeniorhynchus*, shared a similar feeding pattern as *Cx. gelidus*, feeding on humans and bearded pig. One *Cx. quinquefasciatus* fed on wild boar. At this site, three *Culex* spp. fed on humans, two on bearded pigs, four on wild boar, two on red jungle fowl, one on a *Hylarana* frog, and one on a Malaysian field rat. At site R_001, *Cx. quinquefasciatus* fed on both humans and wild boar. At site AG_017, a durian farm, two unidentified *Culex* sp. fed on a buffy fish owl and a human. At secondary forest site SF_017 two *Culex* sp. fed on fairy pitta birds and one on a flying lizard. Last, at secondary forest site F_123 two *Uranotaenia* mosquitoes fed on an amphibian, a spotted-litter frog, and a reptile, a skink.

The second scenario, in which multiple mosquito taxa feed on a single host species, only played out at one site within the rural-built up land cover type site R_013, where wild boar was host for one *Cx. quinquefasciatus* and two *Ae. albopictus* (Table 2).

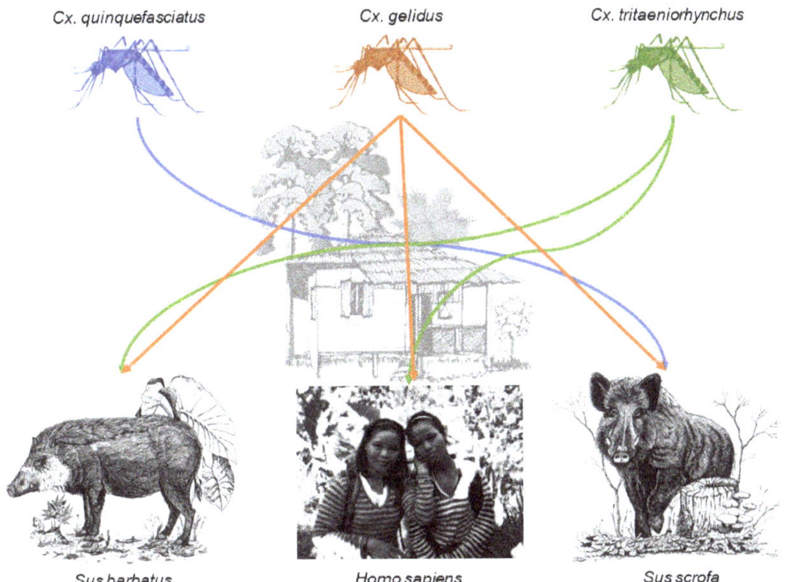

Figure 3. Connections between humans and pig species mediated by designated *Culex* species within the rural built-up site R_015. Samples sizes are found in Table 2. Host species images from left to right are Bornean bearded pig (*Sus barbatus*) [57]†, human (photograph taken by Katherine I. Young), and Eurasian wild boar (Sus scrofa) [58]†. † Licensee of the Licensed Material: Ecology, Conservation and Management of Wild Pigs and Peccaries, Bearded pig *Sus barbatus* (Müller, 1838), Eurasian wild boar *Sus scrofa* (Linnaeus, 1758), License Date: 03/26/2020, PLSclear Ref No.: 35570. Reproduced with permission of The Licensor through PLSclear

Table 2. Mosquito and host networks at the site level in sites where either a single mosquito taxon fed on multiple host taxa or multiple mosquito taxa fed on a single host taxon.

Site ID	Mosquito Taxa	Host Taxa	Number Bloodmeals Identified
		Urban Built-Up	
U_009	*Culex quinquefasciatus*	*Gallus gallus*	1
	Culex quinquefasciatus	*Streptopelia chinensis*	1
		Rural Built-Up	
R_015	*Culex gelidus*	*Homo sapiens*	1
	Culex gelidus	*Sus barbatus*	2
	Culex gelidus	*Sus scrofa*	3
	Culex tritaeniorhynchus	*Homo sapiens*	1
	Culex tritaeniorhynchus	*Sus barbatus*	1
	Culex quinquefasciatus	*Sus scrofa*	1
	Culex species	*Gallus gallus*	2
	Culex species	*Homo sapiens*	3
	Culex species	*Hylarana* species	1
	Culex species	*Rattus tiomanicus*	1
	Culex species	*Sus barbatus*	2
	Culex species	*Sus scrofa*	4
R_001	*Culex quinquefasciatus*	*Gallus gallus*	5
	Culex quinquefasciatus	*Sus scrofa*	1
R_013	*Aedes albopictus*	*Sus scrofa*	2
	Culex quinquefasciatus	*Sus scrofa*	1
		Durian Farm	
Ag_017	*Culex species*	*Ketupa ketupu*	1
	Culex species	*Homo sapiens*	1
		Secondary Forest	
SF_017	*Culex species*	*Draco* species	1
	Culex species	*Pitta nympha*	2
F_123	*Uranotaenia* species	*Leptobrachium hendricksoni*	1
	Uranotaenia species	*Scincidae* family	1

4. Discussion

Our data on mosquito bloodmeal sources from a range of land cover types in Sarawak, Malaysian Borneo demonstrate that bloodmeal analysis from mosquitoes can be used to identify a broad range of hosts in a biodiversity hot spot. In our sample of 134 engorged female mosquitoes, 87% of the bloodmeal sources were identified, similar to other studies of mosquito host feeding patterns [59–62]. Eighty-three percent of the 18 samples that failed to produce host identifications were collected in the KK and OP projects. Mosquitoes from these projects were not held on ice packs during transportation back to the laboratory, while samples from the LC project were held on ice, which may account for the greater success in amplifying samples from the LC project. Storage method and temperature has been shown to impact the success of DNA amplification from mosquito bloodmeals [63,64]. Further, we did not attempt to characterize the digestion status of the bloodmeals collected in any of the projects. DNA degradation of the bloodmeal occurs over time and negatively impacts the success of DNA extraction and bloodmeal identification [63,65,66]. It is also possible that certain host species were not detected due to the specificity of the primers used. A recent study on the Ryukyu Archipelago of Japan reported specialized feeding patterns of *Ae. baisasi* on 15 different fish species determined using universal COI primers and fish specific mitochondrial 12S rRNA primers [51].

Several mosquitoes known to act as key arbovirus vectors were analyzed in this study. Across seven land cover types, humans and boars were the only hosts detected from 16 *Ae. albopictus*, an important vector of human-endemic DENV within this region and a potential bridge vector for sylvatic DENV as well as other arboviruses. Despite its spread into urban areas [67], *Ae. albopictus* is thought to be sylvatic in nature and generalist in feeding behavior [68]. *Ae. albopictus* has been shown to feed on birds, amphibians, and reptiles across its global range, though the majority of bloodmeals characterized come from mammals [60,69]. Within its native range, *Ae. albopictus* seems to preferentially feed on humans, although the majority of studies supporting this conclusion have been conducted in urban and rural land cover types where humans are highly abundant [60,69,70]. However, Sivan et al. reported a reduction in human feeding by *Ae. albopictus* from urban to forested landscapes in India [71]. Our data revealed humans as hosts for *Ae. albopictus* in both secondary and primary forests where this species may play a role as a bridge vector in sylvatic DENV spillover to humans [7,33]. We detected no non-human primates as hosts for *Ae. albopictus*, although Khaklang and Kittayapong reported seven cases in which monkeys and humans were detected from a single bloodmeal from *Ae. albopictus* samples in sites near national parks in Thailand [72]. Further studies on the feeding behavior of *Ae. albopictus* in forests, would be valuable for better determining its role as an arbovirus bridge vector.

Several *Culex* species known to act as vectors of JEV were collected in this study, including *Cx. tritaeniorhynchus*, *Cx. gelidus*, and *Cx. quinquefasciatus* [73]. *Cx. tritaeniorhynchus* has been described as the main vector of JEV transmission in Southeast Asia, where it is highly zoophilic and facilitates amplification of the virus in pigs [6,73,74]. Humans and bearded pigs were the only hosts identified from *Cx. tritaeniorhynchus* from a single site in the rural built-up land cover type. The last study of JEV transmission and feeding habits of *Cx. tritaeniorhynchus* in this region of Sarawak was conducted in the early 1970s, and this species was also found to feed on pigs and humans in rural villages [75]. Based on the bloodmeals taken by other mosquito species, our data show a diversity of hosts available within the rural built-up land cover type; nonetheless, the 2 *Cx. tritaeniorhynchus* sampled selected only humans and pigs. The other JEV vectors mentioned above were also found to feed on Suidae, humans, and in the case of *Cx. quinquefasciatus*, birds, in urban and rural built-up land cover types. The majority of host feeding behavior studies for *Cx. quinquefasciatus* have been conducted in North America and Europe due to its high competence for West Nile virus (WNV) transmission. In this region, this species is an opportunistic feeder and bridge vector [76,77]. Hosts identified from *Cx. quinquefasciatus* across all land cover types were primarily birds, though none of the species documented are known hosts of JEV. *Cx. gelidus*, which previous studies have shown to be an opportunistic feeder [59,78], was only collected in the rural built-up land cover type where it fed on humans and Suidae. Importantly, all engorged *Cx. gelidus* and *Cx. tritaeniorhynchus* were collected from the same site, R_015, where their intertwined host utilization reveals potential risk for JEV transmission. The remaining *Culex* specimens were mostly identified only to genus but showed a wide land cover distribution and breadth of hosts, including the only non-human primate detected in this study, the Horsfield's tarsier. This group of *Culex* also created within-site host networks, the most important of these with *Cx. tritaeniorhynchus* and *Cx. gelidus*, which created connections among six taxa. Identifying these individual species will provide valuable information about their potential vector status, and work to this end is ongoing. We focus on these site-level interactions not to draw general inferences, which is not possible due to the small samples sizes, but rather to offer a model for how this type of data should be parsed to gain maximum insight.

Two other groups of mosquitoes, *Uranotaenia* and a group of 16 individuals identified to the Aedini tribe, made up a significant portion of the bloodmeals identified in this study. *Uranotaenia* species are not considered relevant vectors of human disease, although WNV has been isolated from *Ur. unguiculata* in Eastern Europe [79]. Little is known about this genus of mosquitoes, but some species have been shown to preferentially feed on birds, herptiles, and invertebrate, annelid hosts [61,79]. Some of these feeding preferences were corroborated in our data, extending existing knowledge of the feeding habits of this genus. Humans were the only hosts detected from the Aedini group across multiple swamp forest sites, making this group of individuals the only complete specialists identified in this study. These samples were originally morphologically identified as belonging to

Verrallina, a genus within the Aedini tribe forming a monophyletic group most closely related to *Aedes* [80]. The 95 known species of *Verrallina* primarily occur across the Indomalayan and Australasian biogeographic realms and prefer habitats with flushed pools of brackish water [81–83], similar to the swamp forest sites in which these Aedini specimens were collected. Though *Verrallina* is not considered a medically-important genus of mosquito vectors, they are considered pest species that frequently bite humans [82]. Further, some species have been shown to be experimentally competent for Barmah Forest and Ross River viruses in Australia [84,85]. Though these viruses are not known to occur in Borneo, solid identification of these mosquitoes is necessary to better understand their potential as vectors within this region.

The remaining mosquito taxa collected in this study represent single specimens from secondary forests. Of the five mosquitoes identified to genus, three genera, *Anopheles*, *Armigeres*, and *Mansonia*, include known vectors of human pathogens in this region [86]. Based on sample size, no clear host network information can be visualized for these species. However, the hosts detected from these specimens are diverse, two of which, Bornean bearded pig and fairy pitta, are considered vulnerable species by the International Union for Conservation of Nature (IUCN) [87]. Further, the only non-human primate detected in this study, Horsfield's Tarsier, is also considered vulnerable by IUCN [87] and was a host for an unidentified *Culex* species. Bloodmeals from leeches have been used in Southeast Asia to detect mammalian hosts for conservation efforts [88], and Kocher et al. (2017) took a similar approach in French Guiana with dipteran vectors of disease [89]. This latter study sampled bloodfed Phlebotominae flies and mosquitoes across a gradient of anthropogenic pressure on the local native rainforest forest and compared the diversity of hosts identified. Similar to our data, they found that the diversity of hosts detected was reduced with the increase of anthropogenic disturbance of the surrounding forest [89]. Further, mosquito bloodmeals can be used to give insight into the prevalence of pathogens circulating in wildlife that can pose a significant disease risk and impact conservation efforts [90]. Given the impact of LCLUC on the biodiversity of terrestrial animals in Borneo [91–93] and diversity of hosts detected in our study, a targeted surveillance of mosquitoes and other hematophagous insects is warranted for use in species detections and conservation efforts.

Humans were detected as bloodmeal sources for mosquitoes, many of known vector status, in all land cover types investigated except, ironically, urban built-up. From these data we can conclude that humans are encountered and desired as hosts, despite a relatively high diversity of hosts available across all land cover types, and that there is potential for zoonotic arboviral spillover between land cover types. Future research should focus on arbovirus surveillance in mosquitoes in Sarawak using virus isolation and/or next generation sequencing from which host, vector, and pathogen can be detected. Such surveillance would be most profitable at the interior and border of agricultural, rural, and forested land cover types in Borneo where reservoir hosts occur and humans live in close proximity to reservoir species habitats [1,11,94].

Given that we identified 21 host taxa from a relatively small sample of mosquitoes, it is almost a certainty that more extensive sampling efforts would increase the richness of host species and vector species detected and the complexity of host–vector networks. However, the interactions identified in this study would still hold. Moreover, few studies have performed bloodmeal analyses from mosquitoes in Borneo; to date, only Brown et al. have reported the identification of 38 bloodmeals from mosquitoes in Sabah, Malaysian Borneo [52]. Thus, our data give valuable insight into the vector and host networks related to arbovirus transmission, especially JEV and DENV, which can be used for targeted surveillance of zoonotic arboviruses within this region of Borneo.

Supplementary Materials: The following are available online at http://www.mdpi.com/2414-6366/5/2/51/s1, File S1: mosquito_bloodmeal_identifications.xlsx, File S2: failed_mosquito_amplicons.xlsx, Figure S1: Photographs of land cover types, Table S1: Land cover and site descriptions, Table S2: Bloodmeal PCR primers, Table S3: Bloodmeal PCR reaction information, Table S4: PCR reaction and conditions for cloning.

Author Contributions: Conceptualization, K.I.Y., L.L.C., N.V., D.P, K.A.H.; methodology, K.I.Y., L.L.C., R.J.P, M.B., K.A.H.; data curation, K.I.Y., J.T.M., S.R.A., R.M.H., H.D., M.B.; writing—original draft preparation, K.I.Y., J.T.M, R.H., K.A.H.; writing—review and editing, K.I.Y., J.T.M.,S.R.A.,L.L.C.,H.D.,R.M.H., R.J.P, M.B., N.V., D.P., K.A.H.;

supervision, R.J.P., L.L.C., J.P., N.V., D.P., K.A.H..; project administration, L.L.C., N.V., D.P., K.A.H.; funding acquisition, K.I.Y., H.D., L.L.C., N.V., D.P., K.A.H. All authors have read and agreed to the published version of the manuscript.

Funding: This research was funded by an International Collaborations in Infectious Disease Research (ICIDR) grant from the National Institutes of Health, U01 AI115577 to N.V, D.P, and K.A.H, a pilot grant by the Institute for Human Infections and Immunity to N.V, an AAAS Women's International Research Collaboration Grant to K.A.H. Financial support for H.D. was provided by the Students Training in Advanced Research (STAR) Program through NIH T35 OD010956, and a Fulbright fellowship to K.I.Y.

Acknowledgments: We thank the local communities in Sarawak. Field assistants, Anita O'oi, Miko Anak, and Haziqqah Sani, M.S. Chang for assistance identifying mosquitoes, and The Institute of Biodiversity and Environmental Conservation at UNIMAS, specifically Gabriel Tonga and Alek Tuen for their assistance creating collaborations.

Conflicts of Interest: The authors declare no conflict of interest.

References

1. Weaver, S.C.; Charlier, C.; Vasilakis, N.; Lecuit, M. Zika, Chikungunya, and other emerging vector-borne viral diseases. *Annu. Rev. Med.* **2018**, *69*, 395–408. [CrossRef] [PubMed]
2. Mayer, S.V.; Tesh, R.B.; Vasilakis, N. The emergence of arthropod-borne viral diseases: A global prospective on dengue, chikungunya and zika fevers. *Acta Trop.* **2017**, *166*, 155–163. [CrossRef] [PubMed]
3. Vasilakis, N.; Weaver, S.C. Flavivirus transmission focusing on Zika. *Curr. Opin. Virol.* **2017**, *22*, 30–35. [CrossRef] [PubMed]
4. Weaver, S.C.; Barrett, A.D.T. Transmission cycles, host range, evolution and emergence of arboviral disease. *Nat. Rev. Microbiol.* **2004**, *2*, 789–801. [CrossRef] [PubMed]
5. Coffey, L.L.; Forrester, N.; Tsetsarkin, K.; Vasilakis, N.; Weaver, S.C. Factors shaping the adaptive landscape for arboviruses: Implications for the emergence of disease. *Future Microbiol.* **2013**, *8*, 155–176. [CrossRef] [PubMed]
6. Weaver, S.C.; Reisen, W.K. Present and future arboviral threats. *Antivir. Res.* **2010**, *85*, 328–345. [CrossRef]
7. Vasilakis, N.; Cardosa, J.; Hanley, K.A.; Holmes, E.C.; Weaver, S.C. Fever from the forest: Prospects for the continued emergence of sylvatic dengue virus and its impact on public health. *Nat. Rev. Microbiol.* **2011**, *9*, 532–541. [CrossRef]
8. Pearce, J.C.; Learoyd, T.P.; Langendorf, B.J.; Logan, J.G. Japanese encephalitis: The vectors, ecology and potential for expansion. *J. Travel Med.* **2018**, *25*, S16–S26. [CrossRef]
9. Kent, R.J. Molecular methods for arthropod bloodmeal identification and applications to ecological and vector-borne disease studies. *Mol. Ecol. Resour.* **2009**, *9*, 4–18. [CrossRef]
10. Woolhouse, M.E.; Taylor, L.H.; Haydon, D.T. Population biology of multihost pathogens. *Science* **2001**, *292*, 1109–1112. [CrossRef]
11. Borremans, B.; Faust, C.; Manlove, K.R.; Sokolow, S.H.; Lloyd-Smith, J.O. Cross-species pathogen spillover across ecosystem boundaries: Mechanisms and theory. *Philos. Trans. R. Soc. B Biol. Sci.* **2019**, *374*, 20180344. [CrossRef] [PubMed]
12. Takken, W.; Verhulst, N.O. Host preferences of blood-feeding mosquitoes. *Annu. Rev. Entomol.* **2013**, *58*, 433–453. [CrossRef] [PubMed]
13. Myers, N.; Mittermeier, R.A.; Mittermeier, C.G.; Da Fonseca, G.A.B.; Kent, J. Biodiversity hotspots for conservation priorities. *Nature* **2000**, *403*, 853–858. [CrossRef] [PubMed]
14. Bryan, J.E.; Shearman, P.L.; Asner, G.P.; Knapp, D.E.; Aoro, G.; Lokes, B. Extreme differences in forest degradation in Borneo: Comparing practices in Sarawak, Sabah, and Brunei. *PLoS ONE* **2013**, *8*, 1–7. [CrossRef] [PubMed]
15. Koh, L.P.; Miettinen, J.; Liew, S.C.; Ghazoul, J. Remotely sensed evidence of tropical peatland conversion to oil palm. *Proc. Natl. Acad. Sci. USA* **2011**, *108*, 5127–5132. [CrossRef]
16. Gaveau, D.L.A.; Sheil, D.; Salim, M.A.; Arjasakusuma, S.; Ancrenaz, M.; Pacheco, P.; Meijaard, E. Rapid conversions and avoided deforestation: Examining four decades of industrial plantation expansion in Borneo. *Sci. Rep.* **2016**, *6*, 32017. [CrossRef]

17. Gaveau, D.L.A.; Sloan, S.; Molidena, E.; Yaen, H.; Sheil, D.; Abram, N.K.; Ancrenaz, M.; Nasi, R.; Quinones, M.; Wielaard, N.; et al. Four decades of forest persistence, clearance and logging on Borneo. *PLoS ONE* **2014**, *9*, 1–11. [CrossRef]
18. Miettinen, J.; Shi, C.; Liew, S.C. Land cover distribution in the peatlands of Peninsular Malaysia, Sumatra and Borneo in 2015 with changes since 1990. *Glob. Ecol. Conserv.* **2016**, *6*, 67–78. [CrossRef]
19. Patz, J.A.; Olson, S.H.; Uejio, C.K.; Gibbs, H.K. Disease emergence from global climate and land use change. *Med. Clin. N. Am.* **2008**, *92*, 1473–1491. [CrossRef]
20. Lambin, E.F.; Tran, A.; Vanwambeke, S.O.; Linard, C.; Soti, V. Pathogenic landscapes: Interactions between land, people, disease vectors, and their animal hosts. *Int. J. Health Geogr.* **2010**, *9*, 54. [CrossRef]
21. Rogalski, M.A.; Gowler, C.D.; Shaw, C.L.; Hufbauer, R.A.; Duffy, M.A. Human drivers of ecological and evolutionary dynamics in emerging and disappearing infectious disease systems. *Philos. Trans. R. Soc. Lond. B. Biol. Sci.* **2017**, *372*. [CrossRef] [PubMed]
22. Faust, C.L.; McCallum, H.I.; Bloomfield, L.S.P.; Gottdenker, N.L.; Gillespie, T.R.; Torney, C.J.; Dobson, A.P.; Plowright, R.K. Pathogen spillover during land conversion. *Ecol. Lett.* **2018**, *21*, 471–483. [CrossRef] [PubMed]
23. Platt, G.; Way, H.; Bowen, E.; Simpson, D.; Hill, M. Arbovirus infections in Sarawak, October 1968–February 1970 Tembusu and Sindbis virus isolations from mosquitoes. *Ann. Trop. Med. Parasitol.* **1975**, *69*, 65–71. [CrossRef] [PubMed]
24. How Ooi, M.; Lewthwaite, P.; Foo Lai, B.; Mohan, A.; Clear, D.; Lim, L.; Krishnan, S.; Preston, T.; Hee Chieng, C.; Hooi Tio, P.; et al. The epidemiology, clinical features, and long-term prognosis of Japanese encephalitis in Central Sarawak, Malaysia, 1997–2005. *Clin. Infect. Dis.* **2008**, *47*, 458–468. [CrossRef]
25. Simpson, D.I.H.; Smith, C.E.G.; Bowen, E.T.W.; Platt, G.S.; Way, H.; McMahon, D.; Bright, W.F.; Hill, M.N.; Mahadevan, S.; Macdonald, W.W. Arbovirus infections in Sarawak: Virus isolations from mosquitoes. *Ann. Trop. Med. Parasitol.* **1970**, *64*, 137–151. [CrossRef]
26. Ooi, C.H.; Bujang, M.A.; Tg Abu Bakar Sidik, T.M.I.; Ngui, R.; Lim, Y.A.-L. Over two decades of Plasmodium knowlesi infections in Sarawak: Trend and forecast. *Acta Trop.* **2017**, *176*, 83–90. [CrossRef]
27. Millar, S.B.; Cox-Singh, J. Human infections with Plasmodium knowlesi—Zoonotic malaria. *Clin. Microbiol. Infect.* **2015**, *21*, 640–648. [CrossRef]
28. Liu, W.; Pickering, P.; Duchêne, S.; Holmes, E.C.; Aaskov, J.G. Highly divergent dengue virus type 2 in traveler returning from Borneo to Australia. *Emerg. Infect. Dis.* **2016**, *22*, 2146–2148. [CrossRef]
29. Pyke, A.T.; Moore, P.R.; Taylor, C.T.; Hall-Mendelin, S.; Cameron, J.N.; Hewitson, G.R.; Pukallus, D.S.; Huang, B.; Warrilow, D.; Van den Hurk, A.F. Highly divergent dengue virus type 1 genotype sets a new distance record. *Sci. Rep.* **2016**, *6*, 22356. [CrossRef]
30. Cardosa, J.; Ooi, M.H.; Tio, P.H.; Perera, D.; Holmes, E.C.; Bibi, K.; Manap, Z.A. Dengue virus serotype 2 from a sylvatic lineage isolated from a patient with dengue hemorrhagic fever. *PLoS Negl. Trop. Dis.* **2009**, *3*, 1–5. [CrossRef]
31. Teoh, B.-T.; Sam, S.-S.; Abd-Jamil, J.; AbuBakar, S. Isolation of ancestral sylvatic dengue virus type 1, Malaysia. *Emerg. Infect. Dis.* **2010**, *16*, 1783. [CrossRef] [PubMed]
32. Kottek, M.; Grieser, J.; Beck, C.; Rudolf, B.; Rubel, F. World map of the köppen-geiger climate classification updated. *Meteorol. Z.* **2006**, *15*, 259–263. [CrossRef]
33. Young, K.I.; Mundis, S.; Widen, S.G.; Wood, T.G.; Tesh, R.B.; Cardosa, J.; Vasilakis, N.; Perera, D.; Hanley, K.A. Abundance and distribution of sylvatic dengue virus vectors in three different land cover types in Sarawak, Malaysian Borneo. *Parasites Vectors* **2017**, *10*, 406. [CrossRef] [PubMed]
34. Richter, R.; Schlapfer, D. Atmostpheric/Topographic Correction for Satellite Imagery (ATCOR-2/3 User Guide, Version 9.3.0, April 2019). Available online: https://www.rese-apps.com/pdf/atcor3_manual.pdf (accessed on 16 October 2019).
35. Richter, R. *ATCOR 2 and ATCOR 3 for Satellite Systems*; ReSe Applications Schläpfer: Wil, Switzerland, 2019.
36. *USGS Shuttle Radar Topography Mission, 1 Arc Second Digital Elevation Model*; Earth Resources Observation and Science Center: Sioux Falls, SD, USA, 2000.
37. Rodriguez, E.; Morris, C.S.; Belz, J.E.; Chapin, E.; Martin, J.; Daffer, W.; Hensley, S. An Assessment of the SRTM Topographic Products. Available online: https://www2.jpl.nasa.gov/srtm/SRTM_D31639.pdf (accessed on 16 October 2019).

38. Obenauer, P.J.; Abdel-Dayem, M.S.; Stoops, C.A.; Villinski, J.T.; Tageldin, R.; Fahmy, N.T.; Diclaro, J.W.; Bolay, F. Field responses of *Anopheles gambiae* complex (Diptera: Culicidae) in Liberia using yeast-generated carbon dioxide and synthetic lure-baited light traps. *J. Med. Entomol.* **2013**, *50*, 863–870. [CrossRef] [PubMed]
39. Reuben, R.; Tewari, S.C.; Hiriyan, J.; Akiyama, J. Illustrated keys to species of *Culex* (*Culex*) associated with Japanese encephalitis in Southeast Asia (Diptera: Culicidae). *Mosq. Syst.* **1994**, *26*, 75–96.
40. Huang, Y.-M.; Rueda, L.M. Description of a paralectotype female of *Aedes* (Finlaya) *niveus* (Ludlow)(Diptera: Culicidae). *Proc. Entomol. Soc. Wash.* **1998**, *100*, 824–827.
41. Rueda, L.M. Pictorial keys for the identification of mosquitoes (Diptera: Culicidae) associated with dengue virus transmission. *Zootaxa* **2004**, *589*. [CrossRef]
42. Walter Reed Biosystematics Unit PACOM Arthropod Identification Keys. Available online: http://www.wrbu.org/aors/pacom_Keys.html (accessed on 10 March 2016).
43. Thiemann, T.C.; Brault, A.C.; Ernest, H.B.; Reisen, W.K. Development of a high-throughput microsphere-based molecular assay to identify 15 common bloodmeal hosts of *Culex* mosquitoes. *Mol. Ecol. Resour.* **2012**, *12*, 238–246. [CrossRef]
44. Townzen, J.S.; Brower, A.V.Z.; Judd, D.D. Identification of mosquito bloodmeals using mitochondrial cytochrome oxidase subunit I and cytochrome b gene sequences. *Med. Vet. Entomol.* **2008**, *22*, 386–393. [CrossRef]
45. Hebert, P.D.N.; Cywinska, A.; Ball, S.L.; De Waard, J.R. Biological identifications through DNA barcodes. *Proc. R. Soc. Lond. Ser. B Biol. Sci.* **2003**, *270*, 313–321. [CrossRef]
46. Hebert, P.D.N.; Ratnasingham, S.; De Waard, J.R. Barcoding animal life: Cytochrome *c* oxidase subunit 1 divergences among closely related species. *Proc. R. Soc. Lond. Ser. B Biol. Sci.* **2003**, *270*. [CrossRef] [PubMed]
47. Tizard, J.; Patel, S.; Waugh, J.; Tavares, E.; Bergmann, T.; Gill, B.; Norman, J.; Christidis, L.; Scofield, P.; Haddrath, O.; et al. DNA barcoding a unique avifauna: An important tool for evolution, systematics and conservation. *BMC Evol. Biol.* **2019**, *19*, 52. [CrossRef] [PubMed]
48. Folmer, O.; Black, M.; Hoeh, W.; Lutz, R.; Vrijenhoek, R. DNA primers for amplification of mitochondrial cytochrome c oxidase subunit I from diverse metazoan invertebrates. *Mol. Mar. Biol. Biotechnol.* **1994**, *3*, 294–299. [PubMed]
49. RAWGraphs Team. How to Make An Alluvial Diagram. Available online: https://rawgraphs.io/learning/how-to-make-an-alluvial-diagram/ (accessed on 16 September 2019).
50. Gyawali, N.; Taylor-Robinson, A.W.; Bradbury, R.S.; Huggins, D.W.; Hugo, L.E.; Lowry, K.; Aaskov, J.G. Identification of the source of blood meals in mosquitoes collected from north-eastern Australia. *Parasites Vectors* **2019**, *12*, 198. [CrossRef]
51. Miyake, T.; Aihara, N.; Maeda, K.; Shinzato, C.; Koyanagi, R.; Kobayashi, H.; Yamahira, K. Bloodmeal host identification with inferences to feeding habits of a fish-fed mosquito, *Aedes baisasi*. *Sci. Rep.* **2019**, *9*, 4002. [CrossRef]
52. Brown, R.; Hing, C.T.; Fornace, K.; Ferguson, H.M. Evaluation of resting traps to examine the behaviour and ecology of mosquito vectors in an area of rapidly changing land use in Sabah, Malaysian Borneo. *Parasites Vectors* **2018**, *11*, 346. [CrossRef]
53. Toma, T.; Miyagi, I.; Tamashiro, M. Blood meal identification and feeding habits of *Uranotaenia* species collected in the Ryukyu archipelago. *J. Am. Mosq. Control Assoc.* **2014**, *30*, 215–218. [CrossRef]
54. Kirchgatter, K.; Tubaki, R.M.; Dos Malafronte, R.S.; Alves, I.C.; De Lima, G.F.M.C.; De Guimarães, L.O.; De Zampaulo, R.A.; Wunderlich, G. *Anopheles* (Kerteszia) *cruzii* (Diptera: Culicidae) em área peridomiciliar durante transmissão de malária assintomática na Mata Atlântica: Identificação molecular das fontes de repasto sanguíneo indica humanos como hospedeiros intermediários primários. *Rev. Inst. Med. Trop. Sao Paulo* **2014**, *56*, 403–409. [CrossRef]
55. Abella-Medrano, C.A.; Ibáñez-Bernal, S.; Carbó-Ramírez, P.; Santiago-Alarcon, D. Blood-meal preferences and avian malaria detection in mosquitoes (Diptera: Culicidae) captured at different land use types within a neotropical montane cloud forest matrix. *Parasitol. Int.* **2018**, *67*, 313–320. [CrossRef]
56. Verdonschot, P.F.M.; Besse-Lototskaya, A.A. Flight distance of mosquitoes (Culicidae): A metadata analysis to support the management of barrier zones around rewetted and newly constructed wetlands. *Limnologica* **2014**, *45*, 69–79. [CrossRef]
57. Luskin, M.S.; Ke, A. Bearded pig *Sus barbatus* (Müller, 1838). In *Ecology, Conservation and Management of Wild Pigs and Peccaries*; Cambridge University Press: Cambridge, UK, 2017; pp. 175–183.

58. Keuling, O.; Podgórski, T.; Monaco, A.; Melletti, M.; Merta, D.; Albrycht, M.; Genov, P.V.; Gethöffer, F.; Vetter, S.G.; Jori, F.; et al. Eurasian wild boar *Sus scrofa* (Linnaeus, 1758). In *Ecology, Conservation and Management of Wild Pigs and Peccaries*; Cambridge University Press: Cambridge, UK, 2017; pp. 202–233.
59. Philip Samuel, P.; Arunachalam, N.; Hiriyan, J.; Tyagi, B.K. Host feeding pattern of Japanese encephalitis virus vector mosquitoes (Diptera: Culicidae) from Kuttanadu, Kerala, India. *J. Med. Entomol.* **2008**, *45*, 927–932. [CrossRef]
60. Faraji, A.; Egizi, A.; Fonseca, D.M.; Unlu, I.; Crepeau, T.; Healy, S.P.; Gaugler, R. Comparative host feeding patterns of the Asian tiger mosquito, *Aedes albopictus*, in urban and suburban Northeastern USA and implications for disease transmission. *PLoS Negl. Trop. Dis.* **2014**, *8*, e3037. [CrossRef] [PubMed]
61. Reeves, L.E.; Holderman, C.J.; Blosser, E.M.; Gillett-Kaufman, J.L.; Kawahara, A.Y.; Kaufman, P.E.; Burkett-Cadena, N.D. Identification of *uranotaenia sapphirina* as a specialist of annelids broadens known mosquito host use patterns. *Commun. Biol.* **2018**, *1*, 92. [CrossRef] [PubMed]
62. Blosser, E.M.; Lord, C.C.; Stenn, T.; Acevedo, C.; Hassan, H.K.; Reeves, L.E.; Unnasch, T.R.; Burkett-Cadena, N.D. Environmental drivers of seasonal patterns of host utilization by *Culiseta melanura* (Diptera: Culicidae) in Florida. *J. Med. Entomol.* **2017**, *54*, 1365–1374. [CrossRef] [PubMed]
63. Oshaghi, M.A.; Chavshin, A.R.; Vatandoost, H.; Yaaghoobi, F.; Mohtarami, F.; Noorjah, N. Effects of post-ingestion and physical conditions on PCR amplification of host blood meal DNA in mosquitoes. *Exp. Parasitol.* **2006**, *112*, 232–236. [CrossRef] [PubMed]
64. Reeves, L.E.; Holderman, C.J.; Gillett-Kaufman, J.L.; Kawahara, A.Y.; Kaufman, P.E. Maintenance of host DNA integrity in field-preserved mosquito (Diptera: Culicidae) blood meals for identification by DNA barcoding. *Parasites Vectors* **2016**, *9*. [CrossRef]
65. Martínez-de la Puente, J.; Ruiz, S.; Soriguer, R.; Figuerola, J. Effect of blood meal digestion and DNA extraction protocol on the success of blood meal source determination in the malaria vector *Anopheles atroparvus*. *Malar. J.* **2013**, *12*, 109. [CrossRef]
66. Mukabana, W.R.; Takken, W.; Seda, P.; Killeen, G.F.; Hawley, W.A.; Knols, B.G.J. Extent of digestion affects the success of amplifying human DNA from blood meals of *Anopheles gambiae* (Diptera: Culicidae). *Bull. Entomol. Res.* **2002**, *92*, 233–239. [CrossRef]
67. Kraemer, M.U.G.; Sinka, M.E.; Duda, K.A.; Mylne, A.Q.N.; Shearer, F.M.; Barker, C.M.; Moore, C.G.; Carvalho, R.G.; Coelho, G.E.; Van Bortel, W.; et al. The global distribution of the arbovirus vectors *Aedes aegypti* and *Ae. albopictus*. *Elife* **2015**, *4*, e08347. [CrossRef]
68. Bonizzoni, M.; Gasperi, G.; Chen, X.; James, A.A. The invasive mosquito species *Aedes albopictus*: Current knowledge and future perspectives. *Trends Parasitol.* **2013**, *29*, 460–468. [CrossRef]
69. Kek, R.; Hapuarachchi, H.C.; Chung, C.-Y.; Humaidi, M.B.; Razak, M.A.B.A.; Chiang, S.; Lee, C.; Tan, C.-H.; Yap, G.; Chong, C.-S.; et al. Feeding host range of *Aedes albopictus* (Diptera: Culicidae) demonstrates its opportunistic host-seeking behavior in rural Singapore. *J. Med. Entomol.* **2014**, *51*, 880–884. [CrossRef] [PubMed]
70. Ponlawat, A.; Harrington, L.C. Blood feeding patterns of *Aedes aegypti* and *Aedes albopictus* in Thailand. *J. Med. Entomol.* **2005**, *42*, 844–849. [CrossRef] [PubMed]
71. Sivan, A.; Shriram, A.N.; Sunish, I.P.; Vidhya, P.T. Host-feeding pattern of *Aedes aegypti* and *Aedes albopictus* (Diptera: Culicidae) in heterogeneous landscapes of South Andaman, Andaman and Nicobar Islands, India. *Parasitol. Res.* **2015**, *114*, 3539–3546. [CrossRef] [PubMed]
72. Khaklang, S.; Kittayapong, P. Species composition and blood meal analysis of mosquitoes collected from a tourist island, Koh Chang, Thailand. *J. Vector Ecol.* **2014**, *39*, 448–452. [CrossRef]
73. Sucharit, S.; Surathin, K.; Shrestha, S.R. Vectors of Japanese encephalitis virus (JEV): Species complexes of the vectors. *Southeast Asian J. Trop. Med. Public Health* **1989**, *20*, 611–621.
74. Van den Hurk, A.F.; Ritchie, S.A.; Mackenzie, J.S. Ecology and geographical expansion of Japanese encephalitis virus. *Annu. Rev. Entomol.* **2009**, *54*, 17–35. [CrossRef]
75. Bendell, P.J.E. Japanese encephalitis in Sarawak: Studies on mosquito behaviour in a land Dayak village. *Trans. R. Soc. Trop. Med. Hyg.* **1970**, *64*, 497–502. [CrossRef]
76. Chancey, C.; Grinev, A.; Volkova, E.; Rios, M. The global ecology and epidemiology of west nile virus. *BioMed Res. Int.* **2015**, *2015*, 376230. [CrossRef]

77. Farajollahi, A.; Fonseca, D.M.; Kramer, L.D.; Marm Kilpatrick, A. "Bird biting" mosquitoes and human disease: A review of the role of *Culex pipiens* complex mosquitoes in epidemiology. *Infect. Genet. Evol.* **2011**, *11*, 1577–1585. [CrossRef]
78. Van den Hurk, A.F.; Nisbet, D.J.; Johansen, C.A.; Foley, P.N.; Ritchie, S.A.; Mackenzie, J.S. Japanese encephalitis on Badu Island, Australia: The first isojation of Japanese encephalitis virus from *Culex gelidus* in the Australasian region and the role of mosquito host-feeding patterns in virus transmission cycles. *Trans. R. Soc. Trop. Med. Hyg.* **2001**, *95*, 595–600. [CrossRef]
79. Filatov, S. Little pigeons can carry great messages: Potential distribution and ecology of *Uranotaenia* (Pseudoficalbia) *unguiculata* Edwards, 1913 (Diptera: Culicidae), a lesser-known mosquito species from the Western Palaearctic. *Parasites Vectors* **2017**, *10*, 464. [CrossRef] [PubMed]
80. Reinert, J.F.; Harbach, R.E.; Kitching, I.J. Phylogeny and classification of Aedini (Diptera: Culicidae), based on morphological characters of all life stages. *Zool. J. Linn. Soc.* **2004**, *142*, 289–368. [CrossRef]
81. Ismail, T.N.S.T.; Kassim, N.F.A.; Rahman, A.A.; Yahya, K.; Webb, C.E. Day biting habits of mosquitoes associated with mangrove forests in Kedah, Malaysia. *Trop. Med. Infect. Dis.* **2018**, *3*, 77. [CrossRef] [PubMed]
82. Harbach, R.E. Mosquito Taxonomic Inventory, Culcidae Classification, Verrallina. Available online: http://mosquito-taxonomic-inventory.info/simpletaxonomy/term/6158 (accessed on 3 October 2019).
83. Rajavel, A.R.; Natarajan, R.; Munirathinam, A.; Vaidyanathan, K. The larval habitat of *Verrallina* (*Verrallina*) *lugubris* and chaetotaxy of field-collected larvae. *J. Am. Mosq. Control Assoc.* **2005**, *21*, 96–97. [CrossRef]
84. Jeffery, J.A.L.; Kay, B.H.; Ryan, P.A. Role of *Verrallina funerea* (Diptera: Culicidae) in transmission of Barmah forest virus and Ross River virus in coastal areas of Eastern Australia. *J. Med. Entomol.* **2006**, *43*, 1239–1247. [CrossRef]
85. Webb, C.E.; Doggett, S.L.; Ritchie, S.A.; Russell, R.C. Vector competence of three Australian mosquitoes, *Verrallina carmenti, Verraullina lineata*, and *Mansonia septempunctata* (Diptera: Culicidae), for Ross River virus. *J. Med. Entomol.* **2008**, *45*, 737–740.
86. Harbach, R.E. Mosquito Taxonomic Inventory. Available online: http://mosquito-taxonomic-inventory.info/ (accessed on 15 September 2019).
87. International Union for Conservation of Nature the IUCN Red List of Threatened Species. Available online: http://www.iucnredlist.org (accessed on 16 October 2019).
88. Tessler, M.; Weiskopf, S.R.; Berniker, L.; Hersch, R.; Mccarthy, K.P.; Yu, D.W.; Siddall, M.E. Bloodlines: Mammals, leeches, and conservation in southern Asia. *Syst. Biodivers.* **2018**, *16*, 488–496. [CrossRef]
89. Kocher, A.; De Thoisy, B.; Catzeflis, F.; Valière, S.; Bañuls, A.L.; Murienne, J. iDNA screening: Disease vectors as vertebrate samplers. *Mol. Ecol.* **2017**, *26*, 6478–6486. [CrossRef]
90. Egizi, A.; Martinsen, E.S.; Vuong, H.; Zimmerman, K.I.; Faraji, A.; Fonseca, D.M. Using bloodmeal analysis to assess disease risk to wildlife at the new northern limit of a mosquito species. *Ecohealth* **2018**, *15*, 543–554. [CrossRef]
91. Scriven, S.A.; Gillespie, G.R.; Laimun, S.; Goossens, B. Edge effects of oil palm plantations on tropical anuran communities in Borneo. *Biol. Conserv.* **2018**, *220*, 37–49. [CrossRef]
92. Twining, J.P.; Bernard, H.; Ewers, R.M. Increasing land-use intensity reverses the relative occupancy of two quadrupedal scavengers. *PLoS ONE* **2017**, *12*, e0177143. [CrossRef] [PubMed]
93. Cheyne, S.M.; Sastramidjaja, W.J.; Rayadin, Y.; Macdonald, D.W. Mammalian communities as indicators of disturbance across Indonesian Borneo. *Glob. Ecol. Conserv.* **2016**, *7*, 157–173. [CrossRef]
94. Hawkes, F.M.; Manin, B.O.; Cooper, A.; Daim, S.; Homathevi, R.; Jelip, J.; Husin, T.; Chua, T.H. Vector compositions change across forested to deforested ecotones in emerging areas of zoonotic malaria transmission in Malaysia. *Sci. Rep.* **2019**, *9*, 13312. [CrossRef] [PubMed]

Tropical Medicine and Infectious Disease

Review

Towards a Sustainable One Health Approach to Crimean–Congo Hemorrhagic Fever Prevention: Focus Areas and Gaps in Knowledge

Teresa E. Sorvillo [1,2,*], Sergio E. Rodriguez [2,3,4], Peter Hudson [5], Megan Carey [3,4], Luis L. Rodriguez [6], Christina F. Spiropoulou [2], Brian H. Bird [1,2], Jessica R. Spengler [2] and Dennis A. Bente [3,4]

1 One Health Institute, School of Veterinary Medicine, University of California Davis, 1089 Veterinary Medicine Drive, Davis, CA 95616, USA; bhbird@ucdavis.edu
2 Viral Special Pathogens Branch, Division of High-Consequence Pathogens and Pathology, Centers for Disease Control and Prevention, Atlanta, GA 30333, USA; pze7@cdc.gov (S.E.R.); ccs8@cdc.gov (C.F.S.); wsk7@cdc.gov (J.R.S.)
3 Department of Microbiology & Immunology, University of Texas Medical Branch, Galveston, TX 77555, USA; mncarey@utmb.edu (M.C.); dabente@utmb.edu (D.A.B.)
4 Galveston National Laboratory, University of Texas Medical Branch, Galveston, TX 77555, USA
5 Huck Institutes of the Life Sciences, The Pennsylvania State University, University Park, PA 16802, USA; pjh18@psu.edu
6 Foreign Animal Disease Research Unit, Plum Island Animal Disease Center, Agricultural Research Service, United States Department of Agriculture, Orient Point, NY 11957, USA; luis.rodriguez@usda.gov
* Correspondence: tesorvillo@ucdavis.edu; Tel.: +1-530-752-7526

Received: 2 June 2020; Accepted: 1 July 2020; Published: 7 July 2020

Abstract: Crimean–Congo hemorrhagic fever virus (CCHFV) infection is identified in the 2018 World Health Organization Research and Development Blueprint and the National Institute of Allergy and Infectious Diseases (NIH/NIAID) priority A list due to its high risk to public health and national security. Tick-borne CCHFV is widespread, found in Europe, Asia, Africa, the Middle East, and the Indian subcontinent. It circulates between ticks and several vertebrate hosts without causing overt disease, and thus can be present in areas without being noticed by the public. As a result, the potential for zoonotic spillover from ticks and animals to humans is high. In contrast to other emerging viruses, human-to-human transmission of CCHFV is typically limited; therefore, prevention of spillover events should be prioritized when considering countermeasures. Several factors in the transmission dynamics of CCHFV, including a complex transmission cycle that involves both ticks and vertebrate hosts, lend themselves to a One Health approach for the prevention and control of the disease that are often overlooked by current strategies. Here, we examine critical focus areas to help mitigate CCHFV spillover, including surveillance, risk assessment, and risk reduction strategies concentrated on humans, animals, and ticks; highlight gaps in knowledge; and discuss considerations for a more sustainable One Health approach to disease control.

Keywords: One Health; spillover; animal-human interface; Crimean–Congo hemorrhagic fever; tick-borne virus; outbreak response; surveillance; tick; livestock; risk reduction

---

1. Introduction

Tick-borne Crimean–Congo hemorrhagic fever virus (CCHFV; family *Nairoviridae*; genus *Orthonairovirus*) causes a severe, often fatal zoonotic hemorrhagic fever in humans, and is listed as a World Health Organization (WHO) and National Institute of Allergy and Infectious Diseases (NIH/NIAID) priority disease based on its capacity for person-to-person transmission, high mortality rate of the disease, and a lack of effective therapeutics or vaccines for human or animal use. Viral isolation and/or disease have been reported from more than 30 countries in Africa, Asia, Southeast Europe, and the Middle East. *Hyalomma* species ticks serve as both reservoir and vector of CCHFV; transmission to humans occurs through tick bite, crushing of engorged ticks, or, to a lesser degree, via contact with body fluids of domestic animals or patients with Crimean–Congo hemorrhagic fever (CCHF). Infection can result in a spectrum of disease; case fatality rates in outbreak settings range between 3% and 80%. Severe disease is characterized by a sudden onset of symptoms, such as high fever, headache, myalgia, and petechial rash, frequently followed by a hemorrhagic state and, occasionally, multiorgan failure [1].

Recently, new foci of CCHF have been identified in several parts of the world, including the Balkan countries, southwest Russia, the Middle East, India, and Spain. Potential reasons for the emergence or re-emergence of CCHF include anthropogenic factors, such as changes in agricultural activities, habitat fragmentation, and importation of infected animals and ticks [2,3]. The potential influence of climate change on the spread of the disease has also been suggested [4–6]. Outbreaks have historically been seasonal and sporadic with clusters of few cases. However, recent increases in the number of cases, especially in Turkey and southwest Asia, have demonstrated the imminent public health impact of this re-emerging disease [7,8]. The potential for zoonotic spillover to humans is high due to its wide geographic spread and persistent circulation in nature. In contrast to other emerging viruses, human-to-human transmission of CCHFV after spillover from the animal host or tick is limited, typically resulting in only small clusters of human disease. The importance of spillover for human disease supports prioritizing measures to mitigate these events.

The CCHFV life cycle involves silent transmission between multiple vertebrate hosts (wild and domestic) feeding immature or adult stages of the tick [9], with the absence of overt clinical disease in both hosts and ticks. This complex cycle, in which the hosts and vectors are by themselves highly influenced by environmental parameters, lends itself well to a One Health approach to disease prevention and control (Figure 1). The concept of One Health recognizes the interconnectedness of human, animal, and environmental health; it is an approach to disease prevention that endeavors to address human health in a broader context, making change and intervening in an interdisciplinary manner [10].

In 2019, a One Health-based framework entitled "A Tripartite Guide to Addressing Zoonotic Diseases in Countries" was published by The Food and Agriculture Organization of the United Nations (fao), the World Organization for Animal Health (oie), and WHO [11]. An important section of the guide focuses on taking a multisectoral One Health approach to disease control, including: (i) strategic planning and emergency preparedness; (ii) surveillance for zoonotic diseases and information sharing; (iii) coordinated investigation and response; (iv) joint risk assessment for zoonotic disease threats; (v) risk reduction, risk communication, and community engagement; and (vi) workforce development. Here, we examine critical topics for CCHF prevention, including surveillance, risk assessment, and risk reduction strategies for humans, animals, and ticks that could lead to a more sustainable One Health approach to disease control and outbreak response. In addition, by discussing several of the parameters and conditions that mitigate or exacerbate virus transmission, we highlight gaps in knowledge that, when addressed, would further support effective One Health control practices.

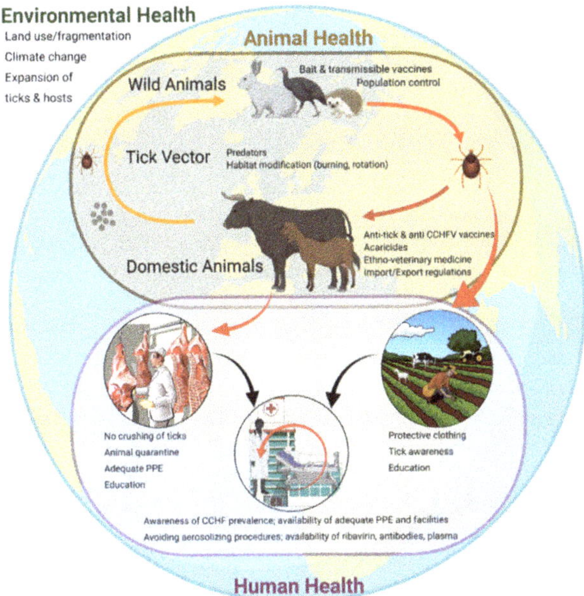

Figure 1. Crimean-Congo hemorrhagic fever virus (CCHFV) routes of transmission and associated intervention measures within the three realms of One Health. Environmental health (outer circle), including climate change and land use, influences several aspects of CCHFV transmission, including tick and animal host distribution and density; tick–animal host interactions affect the overall animal health status and livestock production levels (brown circle). The virus transmission pressure between ticks and animals increases progressively with the life cycle of the tick (reflected by the color gradient of the arrows from light orange to dark orange to red). Human health (purple circle) overlaps with the animal health realm through virus transmission routes, most commonly through tick bites or crushing of ticks, and secondarily through contact with viremic animal blood during the slaughtering process. Human-to-human transmission can also occur in a household or nosocomial setting when no proper personal protective equipment is worn.

2. Surveillance

An effective surveillance system provides data on disease incidence and prevalence, establishes high-risk behaviors or practices at vector/host/human interfaces, and guides the development and implementation of preventive measures. Given the expansive geographic range of *Hyalomma* ticks, their multi-staged life cycle, and corresponding variability in host preferences, surveillance efforts must be coordinated across multiple sectors to inform on CCHFV in humans, animals, and ticks [2,12,13]. International surveillance for CCHF in humans comprises national surveillance programs in coordination with WHO regional offices, collaborating centers, and laboratories. Supplementing WHO-based surveillance networks are other organizations working in both the human and animal health sectors, including the FAO and OIE; non-governmental agencies, such as Médecins Sans Frontières and the Wildlife Conservation Society; and academic partners [14].

Since 2002, an increase in CCHF cases in Turkey resulted in greater awareness of the virus, leading to an influx of serology- and genome-based surveillance studies; over 35% of the CCHF publications in the last 15 years have focused on antibody and genome detection in humans, animals, and ticks (unpublished literature review by the authors). Despite substantial increases in data, many challenges in surveillance efforts remain, including the lack of a framework for extrapolating these data to formulate risk assessments in human populations. Below, we review international and country-specific approaches to surveillance in human, animal, and tick populations. We discuss current strategies and highlight remaining gaps in surveillance systems and methodologies. Detailed information on characteristics of specific diagnostic assays for CCHFV can be found in the WHO Roadmaps meeting report [15].

2.1. Human Surveillance

Five levels were recently proposed to reflect country-specific CCHFV surveillance systems [16]. Levels are based on the incidence of cases, potential for disease transmission to humans, and presence of surveillance systems. Level 1 countries are those in which human CCHF cases are reported annually and the virus is endemic; Level 2 countries have sporadic autochthonous human cases; Level 3 and 4 countries have no documented human cases but ecologic data, including the presence of *Hyalomma* ticks, suggests that cases may occur without detection; and Level 5 countries are ones for which no information is available. Human CCHFV surveillance strategies inherently vary between different country-specific levels. For example, Level 1 and 2 countries require a robust strategy, as human cases can be sporadic and thus public awareness may be lacking. In addition, these countries could have a relatively higher incidence of mild or asymptomatic human infections, which may occur more frequently than previously believed [17–19]. In contrast, Level 5 countries should focus surveillance specifically on the presence of the tick vector and not initially on detecting human seropositives [16].

While surveillance strategies will understandably vary based on the "level" of CCHFV in a country, wide gaps in the strength of human surveillance systems between countries with similar levels of endemicity remain [20]. Even across regions, surveillance approaches tend to lack standardization; their focus may be on acute outbreak-specific response, targeting at-risk populations, or examining larger groups of the population. Nasirian et al. (2019) recently performed a meta-analysis of 45 published human serosurveys, determining that animal contact, husbandry, and farming, as well as tick bites, hunting, and slaughtering were the most evident risk factors for CCHF seropositivity [20]. This report serves as an example of the power of collective analysis of national and international reports; however, the utility of this approach would be greatly enhanced with improved standardization of data collection and reporting.

Although human serosurveillance can be used to generate these types of high-level risk assessments, data should be interpreted with caution. Important gaps in CCHF knowledge are the discrepancy between seroprevalence and symptomatic disease incidence and an understanding of CCHFV strain-specific virulence. More specifically, we find that seropositivity in human populations cannot be assumed to definitively correlate with a risk for human CCHF disease. For example, within Greece, CCHFV was first isolated from a *Rhipicephalus bursa* tick in 1975 (strain AP92/P7) [21]. Subsequent human surveillance data have indicated widely varying rates of seropositivity in different regions within Greece, some as high as 27.5%, yet only one recorded human case of CCHF has been documented within the country since 1975 [22]. It has been hypothesized that the Greek AP92 strain is an outlier, confined only to Greece and avirulent or less virulent than other strains circulating worldwide, though laboratory studies have never been conducted to confirm these hypotheses. However, in recent years, AP92-like strains have been discovered outside of Greece, and interestingly, these genetically similar viruses have been implicated in causing severe human disease [23,24].

Adding to disparate surveillance networks and methodologies are unknowns regarding the accuracy and cross-reactivity of serologic testing. No standard serologic assay has been approved for human clinical use, and the assays used for research purposes have shown variation in performance based on genetic heterogeneity of the virus [25–27]. Thus, human surveillance data alone cannot be used to understand risk in a given geographic area. Rather, a more comprehensive One Health-based approach to CCHF surveillance should be used to provide a more refined understanding of virus spillover risk.

2.2. Animal Surveillance

An extensive number of serosurveys have been performed in domestic and wildlife species informing on both the geographic range and relative levels of CCHFV circulation [28,29]. Animals are fundamental to virus maintenance by serving as hosts for the ticks; they develop a subclinical infection with a relatively low-level viremia typically lasting a week and up to 14 days [30]. The duration of viral replication in other tissues over time has not been characterized in livestock and wild-life species. However, in mouse models of disease, RNA levels in survivors indicate that virus may be detected in certain tissues longer than in blood [31,32], indicating a need for both tissue tropism and viral persistence studies in livestock and other animals.

Serological data are the mainstay in animal surveillance for CCHFV [28]. PCR is of limited value; infectious virus is found only briefly in animals as they have a short period of viremia, and while virus may persist longer in tissues, this period is not thought to be extensive. This brief persistence is reflected in the small number of reported virus isolates obtained directly from animals and the large sample size required to obtain virus isolates [29,33–35]. As such, PCR has limited sensitivity for identifying animals involved in outbreaks or maintenance cycles, even if an exposure event was relatively recent. In contrast, while serology cannot indicate active infection, it can identify past exposure and is not affected by the sensitivity limitations inherent in PCR analyses for CCHFV in animals.

Serological data can be very useful for revealing the geographic range of CCHFV but have limitations. Critical details (e.g., species, breed, age, proportion of herd sampled, sampling dates, etc.) are often lacking, necessitating more standardized methodology and reporting. As with human data, serology in animals must also be interpreted with caution. First, without repeated sampling, these data do not indicate the timing of exposure, emphasizing the importance of longitudinal sampling in a region. While human data support long-lived antibody and CD8+ T-cell-mediated responses [36–38], the antibody responses to virus in livestock (and many other species) have not yet been characterized over time. In experimentally infected mice, anti-CCHFV antibodies generally developed around one week post inoculation (p.i.); IgM appeared, beginning at 8 days p.i. and declined thereafter, whereas IgG levels increased and were sustained at least 9 weeks p.i. Only marginally neutralizing antibody responses were detected by day 28 p.i., and did not substantially increase by 9.5 weeks p.i. [31]. Adding to the complexity, wildlife and livestock are most likely exposed to infected ticks multiple times throughout a tick season and their lifespan. It is unknown how this influences isotype-specific antibody levels over time. Longitudinal sampling in individual animals may be prohibitive from a surveillance standpoint, but experimental data on antibody persistence can inform the frequency of sampling decisions based on the level of animal turnover in the area.

The movement of animals can also complicate serological data interpretation, as the location in which the sample was collected may differ from the location of exposure. Furthermore, maternal transfer of anti-CCHFV antibodies has been demonstrated in sheep [39], suggesting that even if offspring have not moved, the geographic history of the dam may also have to be examined. Finally, cross-reactivity to other nairoviruses should also be considered, as sequence homology has been described between members of different serogroups, including the Nairobi sheep disease serogroup, known to circulate and cause disease in domestic animals [40,41]. Cross-reactivity between nairoviruses has also been reported in protein expression assays [42]. To date, the degree of cross-reactivity between nairoviruses has not been systematically evaluated; evidence suggests that it may occur, but the frequency and level of cross-reactivity are unknown.

2.3. Tick Surveillance

Ticks are the vector and reservoir of CCHFV, maintaining the virus in nature through transstadial and transovarial transmission [29,43–46]. Ticks within the genus *Hyalomma* (family Ixodidae) are considered to be the primary vectors and exhibit a two-host lifecycle, in which ticks in larval and nymphal stages feed on small mammalian hosts and later switch to a different larger mammalian host as adults to complete their life cycle [1]. Evidence has demonstrated that the geographic distribution of *Hyalomma* ticks specifically overlaps with human CCHF cases, which is not the case for other tick species, suggesting that the presence of *Hyalomma* ticks may be necessary to support natural circulation of the virus [29,47–49]. While the role of other tick species in CCHFV maintenance and transmission is not well defined, the virus has been detected in several other tick genera (*Amblyomma*, *Rhipicephalus*, *Dermacentor*, and *Ixodes*); however, data on the vector competence of these species is largely missing [47]. Assessing vector competence is crucial to understanding the potential for CCHFV establishment in new geographic regions, particularly those where non-*Hyalomma* species might supplement the transmission cycle or play a role in cryptic transmission [50].

Hyalomma ticks are broadly distributed across Asia, Europe, and Africa, making the potential for CCHFV circulation and spillover a widespread global concern. Additionally, changing climatic conditions and the frequent importation of *Hyalomma* ticks into novel regions via migratory birds or animal trade (e.g., livestock) is of particular concern for the expansion of endemic areas [51–54]. The detection of viral antigen or genomic fragments in tick vectors has historically been an important precedent and/or predictor of human cases. For example, in 2010, prior to any reports of human cases in Spain, CCHFV RNA was detected in that country in *Hyalomma lusitanicum* and *Hyalomma marginatum* ticks at a rate of 2.78% (44/1579), which was similar to the CCHFV tick infection rate of other endemic countries in Europe, including Kosovo, Bulgaria, and Albania [3,55,56]. These data were important in confirming the established spread of CCHFV into western Europe 7 years before the first human case was documented in Spain in 2017 [55,56]. Additionally, viral antigen was detected in ticks in Iran over 20 years before autochthonous human cases were described [57]. Studies that focus on tick collection from cattle can be a particularly sensitive indicator of the presence of viral circulation in a given geographic area because cattle can be infested by a greater density of *Hyalomma* spp. ticks than small ruminants [58].

Although baseline detection of virus in ticks is important, many other factors that we do not fully understand determine the long-term maintenance and persistence of the virus in ticks over time, including rates of transstadial and transovarial transmission, co-feeding behaviors [59], and relative densities of both ticks and animal hosts. In order to better understand the frequency and contribution of transstadial/transovarial transmission in the maintenance of CCHFV in tick populations, surveillance efforts should focus on collecting questing ticks, though this can be significantly more difficult than collecting ticks from animal hosts. Recently, a growing number of publications have reported the detection of CCHFV in non-*Hyalomma* tick species collected from animal hosts, suggesting that these species may play a role in transmission to humans [47]. It must be emphasized that the detection of virus in non-questing ticks, particularly those collected from animal hosts, cannot be correlated with vector competence or risk for transmission to humans. Vector competence ultimately needs to be evaluated and confirmed in a laboratory setting.

3. Risk Assessment

Although many of the individual determinants leading to spillover, such as virus prevalence in ticks or animals, have been studied extensively, each is typically addressed in isolation within its own discipline. As a result, prevention is often focused on interventions targeting specific aspects of transmission without accounting for other interconnected parameters. This can lead to unproductive intervention efforts (as described in Section 6.3), such as with the release of helmeted guineafowl [60]. A comprehensive understanding of virus maintenance in nature and spillover into human populations, and how these processes are hierarchically, functionally, and quantitatively linked remains a fundamental deficit in CCHF research. Plowright et al. (2017) proposed a mechanistic structure of determinants of disease spillover and their interactions [61]. We propose that this comprehensive framework be adopted to guide research efforts in support of a One Health-based approach for CCHF (Figure 2). In addition to considerations regarding epidemiology, ecology, virology, and vector biology (covered in other sections of this manuscript), mathematical modeling will play an important role in adopting the framework to develop comprehensive risk assessments. Modeling within the context of the framework will serve to highlight key gaps in knowledge, which will facilitate prioritization of epidemiology and laboratory-based experiments and of mitigation policies.

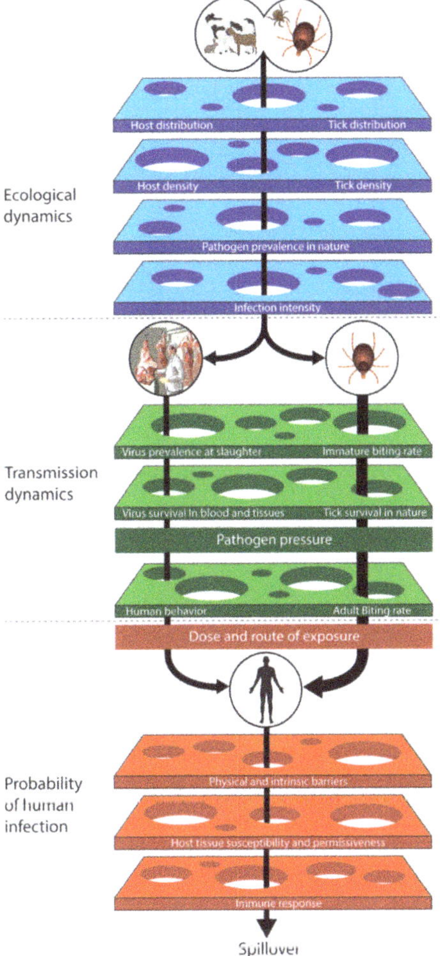

Figure 2. Spillover percolation model of different barriers to CCHFV spillover and opportunities to intervene (based on Plowright et al., 2017) [61]. The risk of CCHFV spillover is determined by the ecological dynamics of infection in the tick and animal host (blue barriers), human epidemiology/behavior (green barriers), and within-host biological factors (orange barriers) that influence exposure outcome in humans. CCHFV must overcome all of the barriers before spillover can occur; thus, each barrier offers intervention opportunities to prevent human disease. The distribution and intensity of infection (virus levels) in tick vectors and animal hosts, virus release from ticks and animal hosts, and CCHFV and tick survival determine the pathogen pressure (the amount of virus that is available to humans at a given point in time and space). Pathogen pressure then interacts with human behavior and the tick vector to determine the likelihood, dose, and route of human exposure. The thickness of the arrows represents the likelihood of spillover to humans (tick > livestock-associated exposure). The probability of human infection is determined by the within-host factors. Once spillover has occurred, human-to-human transmission in a household or nosocomial setting can occur as well (not depicted). In this case, barrier parameters, such as virus dose, route, and human behavior, still apply.

3.1. Modeling Tick-Borne Viruses

Modeling of tick-borne viruses has historically been lacking, primarily due to the complexity of the vector-host-virus system. Mathematical models can be used to integrate virus, vector, and host biology in order to identify key features that increase the risk of infection. These models are little more than a formal description of our biological understanding, but they reveal the non-linearities (a lack of a direct relationship between an independent variable and a dependent variable) associated with transmission and identify major shifts in transmission dynamics that can result in unexpected outbreaks. Working from the tick life cycle and adding key features of CCHFV biology, we can start to identify critical data needed to estimate outbreak risk; we also determine whether an understanding of the system in one location helps to explain the findings in another, and what aspects of the system shape transmission. Ultimately, we seek to obtain insights into the dynamics of the virus in relation to the vector and hosts, which will lead us to identify the type of surveillance that can provide critical information for developing targeted intervention strategies to reduce infection risk in humans and livestock.

Tick models are fundamentally different from mosquito models because ticks bite once per developmental stage, live over multiple seasons, and frequently feed on multiple host species, some of which are capable of transmitting the virus on to other ticks (amplifying hosts) while others are not. There are multiple non-linearities in the system for which the output does not change in direct proportion to a change in the input: The rate of transmission rises as a consequence of host–tick combinations and the resulting heterogeneity of how CCHFV is amplified in this system, leading to transmission and outbreaks driven by certain vector-host combinations. In this respect, models of tick-borne infections are probably best used to determine if our understanding of the biology explains the changes in seroprevalence from year to year in different host species and variation between locations. Additionally, models can be useful in differentiating competing hypotheses and validating models by making predictions that can be tested with critical experiments (Figure 2). Indeed, such experiments frequently provide important data to improve parameter estimation (i.e., finding the unknown parameters of the model that give the best fit to a set of experimental data), and model precision and accuracy.

In some types of analysis, multiple spillover determinants are aggregated, obscuring the interactions between them. Although aggregation might be appropriate at times, sometimes discrete mechanisms can only be identified using simpler models. Non-linear interaction between the spillover determinants and barriers can create bottlenecks that provide opportunities for One Health interventions.

3.2. What Key Data Are Needed?

In endemic areas, seroprevalence in hosts is most likely in an equilibrium and is roughly constant from year to year since livestock hosts are repeatedly exposed to the virus through ticks in various life stages persistently infected with CCHFV. Young CCHFV-naïve animals join the population as old animals die. Therefore, examining seroprevalence across the age curve over years can be some of the most helpful data. This curve quite simply plots the proportion of hosts within each age category that have been exposed to infection and tracks the changes in this proportion with host age. With directly transmitted freely mixed populations, these data can be used to obtain an estimate of the basic reproduction number, R_0, which would be $1 + (L/A)$, where L is the life expectancy of the host and A is the average age of infection [62]. R_0 is useful when there is no previous exposure in the population and when comparing different studies, and is defined as the number of secondary cases in a population of susceptible hosts that arises following the introduction of an infectious individual. R_0 for many vector-borne systems is not simple to estimate but provides a rough approximation that can be very useful for comparing populations and then validating with a more precise metric.

Cooper (2007) applied a basic tick-protozoan model of *Theileria parva* developed by Medley et al. (1993) to CCHF disease dynamics between ticks and their domestic animal hosts [63,64], and clearly showed how seroprevalence data can be used to estimate the rate at which animals become infected.

In this situation, the most useful study would be long-term longitudinal sampling of livestock of a known age so that the time of seroconversion can be cleanly estimated. In the absence of longitudinal studies, detailed sampling across the age structure at one time point can provide interesting data for comparison between sites, assuming the seroprevalence has reached equilibrium.

Ak et al. (2018) developed a computational framework based on Gaussian processes to spatiotemporally predict human cases in Turkey. The predictive value of the framework was tested against CCHF cases in the years 2004 and 2015 [65]. This Gaussian process framework obtained better results than two frequently used standard machine learning algorithms (i.e., random forests and boosted regression trees) under temporal, spatial, and spatiotemporal prediction scenarios. This shows that such frameworks have the potential to make an important contribution to public health policy makers.

3.3. *Important Features of CCHF Modeling*

Several features of the CCHF-tick-host system could generate non-linear responses in transmission and result in disease outbreaks. Here, we mention a few, although there is a pressing need to understand how some of these processes could produce rapid changes in exposure risk. In several instances in which data regarding *Hyalomma* spp. were unavailable, we consider what is known about other ixodid ticks and tick-borne diseases. We acknowledge that translating information from other tick or tick-borne diseases may not be an accurate approach, since the tick and pathogen life cycle are significantly different from those involved in maintaining CCHFV, but it may be a first step in helping frame ideas for the control of CCHF.

3.3.1. Multiple Host Species and the Dilution Effect

Modeled by Norman et al. (1999) [66], the dilution effect occurs when the transmission cycle, maintained by one competent vertebrate host species is reduced by the presence of a second host species that is less competent. The dilution effect has been demonstrated for Lyme disease but is still a controversial hypothesis for other tick-borne diseases. Adding a second host species to the community of ticks effectively increases the number of vertebrate hosts available to feeding ticks so the tick population expands. If this second vertebrate host is less competent for the virus, the virus is lost from the system: The presence of the second host dilutes the virus. This is because ticks feed a limited number of times during their life cycle, and feeding on the less competent host effectively means reduced virus transmission. This results in fewer infectious vertebrates within the population and a lower R_0 value for the virus in that host community, even though the R_0 for the ticks is effectively increased. Field studies on louping ill virus have shown that removing the amplifying host (in this instance, a non-viremic host) results in declining viral abundance in line with the theoretical models [67]. Excellent studies have also been undertaken on the dilution effect of Lyme disease [68].

3.3.2. Non-Viremic Transmission through Co-Feeding

In the case of most vector-borne viruses, transmission occurs when the arthropod vector bites a viremic host and becomes infected. However, some hosts permit direct passage of CCHFV between co-feeding ticks without the requirement of a viremic response from the host. During feeding, ticks produce pheromones that attract other ticks to the same feeding pool, facilitating tick-to-tick transmission, a process accelerated by the presence of tick saliva [69]. The likelihood of co-feeding increases with the intensity of tick infestation, and hosts infested with high numbers of tick larvae and nymphs are more likely to have co-feeding groups. The relative importance of co-feeding transmission in wildlife hosts preferred by *Hyalomma*, like *Leporidae*, is not yet known [70]. Thus, more field studies looking at the densities of *Hyalomma* species ticks on wildlife hosts need to be conducted (Figure 2). The co-feeding transmission phenomenon has been well characterized for tick-borne encephalitis virus (family Flaviviridae; genus *Flavivirus*) and *Ixodes* spp. ticks, yet only one publication addresses this question for CCHFV. A small proportion of ticks (0.1–1.9%) were shown in a laboratory study to acquire CCHFV when co-feeding on non-viremic guinea pigs [70]. However, guinea pigs are not the natural host for *Hyalomma* ticks, nor do they play a role in the natural transmission of the virus. Thus, co-feeding transmission between ticks must be much better characterized in studies with natural hosts in a controlled laboratory setting (Figure 2).

3.3.3. Transovarial Transmission

Passing the virus from the ovaries of female ticks to the egg and eventually the larva is the mechanism for maintaining viral persistence from one generation to the next. Indeed, infecting a high proportion of larval ticks can also result in increased R_0. This has been recorded for CCHFV in *Hyalomma truncatum*, although at a fairly low level [71]. The role of transovarial transmission in different *Hyalomma* species is unclear and needs to be investigated in a controlled laboratory environment (Figure 2).

3.3.4. Sexual Transmission of the Virus in Ticks

Gonzales et al. (1992) showed sexual transmission of CCHFV between copulating *Hyalomma truncatum* ticks, although the sample size was insufficient to reliably estimate the rate of transmission. Since virus-induced mortality is unlikely in ticks, the prevalence of infection is expected to rise from one tick life stage to the next, so sexual transmission is a route by which host infection and transmission of CCHFV to the eggs can increase [71].

3.3.5. Role of Other Tick Species: Vector Competence and Vectorial Capacity

Current knowledge indicates that *Hyalomma* ticks are the primary vector of CCHFV transmission, but in many locations, other tick species are present and may contribute to transmission. Vector competence (the ability to become infected with and transmit a virus) of ticks needs to be tested further to improve our understanding of which species and life stages are important in CCHFV ecology [47]. Besides vector competence, we must also understand vectorial capacity, which is the amount of transmission that occurs. For example, a highly competent vector that rarely feeds on infected hosts or has an extrinsic incubation period (time from virus ingestion to transmission) longer than the time between feedings is unlikely to cause much onward transmission. In contrast, a poorly competent tick vector can initiate an outbreak if it prefers to feed on competent mammalian hosts and has an extrinsic incubation period shorter than the time between feedings. Furthermore, the extrinsic incubation period depends on molecular interactions within the tick, such as the microbiome of the tick gut, cross-immunity with other viruses, and the tick immune response [72], as well as external factors like co-feeding responses and interactions with host susceptibility. Understanding vector competence and vectorial capacity is vital in predicting CCHFV expansion to new geographic areas and must be investigated further (Figure 2).

4. Risk Reduction: Human-Targeted Approaches

4.1. Physical and Chemical

Occupations at high risk for CCHF include veterinarians, abattoir workers, and farmers. For veterinarians and abattoir workers, the use of standard infection control practices when handling potentially infectious blood or ticks aids in risk reduction [73]. Farmers accounted for almost 90% of CCHF cases during a recent outbreak in Turkey [73], emphasizing environmental exposure to ticks as a major source of virus transmission. Basic methods to reduce tick bites, such as wearing appropriate clothing (e.g., long sleeves, pants, etc.), chemical approaches (e.g., repellent), and visual inspection of skin and clothing will contribute to decreased transmission risk. It is noteworthy that few tick repellents have been definitively evaluated for use against *Hyalomma* ticks, and that *Hyalomma* adults have been shown to have a greater affinity for human hosts than other species [29,74–77], potentially reducing the relative efficacy of these approaches for CCHF disease control compared to other tick-borne pathogens.

When attached ticks are found, proper tick removal techniques should be used; improperly removing attached ticks can crush them, spilling infected blood and facilitating virus transmission [78]. Thus, effective risk communication and community education promoting proper tick inspection and removal techniques are vital. The most effective methods for safely removing the whole tick including the mouthparts, while also minimizing the risk of tick-borne disease transmission, are mechanical and can be performed with tools readily available in most regions [78,79]. Specifically, mechanical removal with tweezers (82.5% success rate) was found to be superior to both lassoing (47.5% success rate) and card detachment (7.5% success rate) [79].

Other considerations for tick transmission of CCHFV are: (i) the amount of virus transmitted; (ii) time to transmission post attachment; and (iii) potential enhancement of infection by tick saliva. Unlike other pathogens (e.g., *Borrelia* spp. [80]), the kinetics of CCHFV transmission by *Hyalomma* ticks are not known. Ebel and Kramer (2004) demonstrated that Powassan virus (family Flaviviridae; genus *Flavivirus*) could infect mice within 15 min of tick attachment [81]; this has shaped the dogma that all tick-borne viruses are transmitted immediately after tick attachment. However, a study in mice with Dugbe virus (genus *Orthonairovirus*) demonstrated the absence of viral antigen in the salivary glands of newly attached *Amblyomma* ticks [82]. This indicates that virus transmission in ticks may not always occur as rapidly as suggested in earlier Powassan virus studies. Studies are needed to characterize CCHFV-specific transmission kinetics. However, regardless of the exact timing of virus transmission, prompt tick removal must be emphasized to minimize the risk of transmission.

4.2. Behavior

Human–animal interactions differ across cultures depending on the species of animals present, views towards domestic animals and ticks, and cultural perceptions as a whole. The nature of these interactions may result in activities that increase the risk of CCHFV exposure. Social and cultural practices, particularly those that include the movement of potentially infected animals, have been epidemiologically linked to CCHF outbreaks [83–85]. For example, the practice of livestock (cattle, sheep, goat, or camel) sacrifice central to festivals like the Hajj and Eid-al-Adha results in the contact of large numbers of people with potentially infectious animal blood and body fluids. During Eid-al-Adha, nearly 8 million animals are sacrificed each year in Pakistan alone [84]; 2 million small animals and 750,000 cattle are slaughtered in Turkey, accounting for 25% of all annual slaughtering in that country [84].

The risk of exposure to pathogens like CCHFV during slaughtering is known, and practices have been implemented to reduce exposure during these festivals. In general, individuals performing the sacrificial rite during the Hajj or Eid-al-Adha should adhere to careful techniques, including slaughter at designated sites and sanitary collection and disposal of blood and tissues [9]. In Iran, viral hemorrhagic fever surveillance protocols are in place and report various slaughtering statistics, including the location of slaughtering (i.e., by industrialized, smaller slaughterhouses or by individuals). During Eid, cities coordinate with the Iran Veterinary Organization and create special temporary slaughtering centers. The Saudi Ministry of Health actively discourages amateur slaughtering by encouraging the use of "ritual coupons", allowing the Hajjees to fulfill their sacrificial obligation by proxy without performing the act themselves. Instead, the Saudi Project for Utilization of Hajj Meat performs the ritual with the supervision of government and veterinary workers and then distributes the meat to the poor according to the Muslim tradition. This process could be extended to further mitigate virus transmission to the public. Furthermore, if Hajj authorities banned importation of sacrificial animals from CCHF-endemic countries and instead mandated ritual coupons for travelers, there would be added protection against virus transmission to those making the Hajj. With this plan in place, Muslims from CCHF-endemic countries could still faithfully fulfill their Hajj obligation while minimizing the risk of spreading the virus to non-endemic regions and reducing the chances of an outbreak in Saudi Arabia. Acaricides and quarantine (14 days) are another option for reducing CCHFV transmission during these festivals [83,86]. If acaricides are unavailable or if rates of resistance are high, a mandatory

4-week quarantine for all animals imported for slaughter from CCHF-endemic countries can help mitigate exposure risk. The first 2 weeks would allow any attached ticks to drop off the animal, and the second 2 weeks would allow any active viremia to subside [83].

4.3. Vaccination

A WHO Research and Development Blueprint for Action to Prevent Epidemics working group established a draft roadmap analysis for the CCHF field and outlined a timeline for both benchmarks and deployment goals for a human vaccine [15]. They supported refining the top contending experimental vaccines with a target product profile and promoted clinical safety and early immunogenicity trial testing of selected vaccines by 2019, with phase II trials encouraged for 2023. The draft roadmap analysis prioritizes further animal model development to establish correlates of protection, which can satisfy regulatory authorities for use in pre-clinical evaluations and regulatory approval for human CCHFV vaccines [15].

To date, several CCHFV vaccine candidates have been developed [87,88] with a variety of antigenic variations (strain and/or gene combinations). Vaccine evaluation has been performed using different dosing schemes that involve diverse adjuvants, inoculation routes, challenge strains, and prime/boosts strategies [88,89]. In animal models, vaccination has been shown to generate humoral immunity and can generate up to 100% protective efficacy depending on the platform and approach to vaccination. For the last decade, disease models of CCHFV have been limited to immunocompromised mouse strains [89,90]. The most effective antigen component(s) and correlates of protection for CCHFV vaccines are not yet clear. When used in experimental vaccines, the nucleoprotein generates a more T-cell based response, while other viral antigens, such as whole or portions of the glycoprotein precursor, can generate potent immunoglobulin responses [88–90]. Passive transfer of sera and T-cells generated to either the nucleoprotein or glycoprotein molecules also did not protect but delayed time to death [91–93]. Other studies have implicated that antibodies targeting certain portions of the glycoprotein neutralized CCHFV but also failed to provide protection in vivo [94,95]. Conversely, antibodies targeting non-structural elements of the glycoprotein precursor can produce non-neutralizing protection in animal models [94–96].

To our knowledge, only one study has examined human humoral protective responses to CCHFV vaccination, demonstrating antibody induction with low neutralizing capacity and a T-cell-induced immunity after multiple booster doses [97]. Taken together, the data indicate that experimental vaccines eliciting both robust T-cell-driven responses to nucleoproteins and glycoproteins, in combination with targeted serological responses to non-structural elements of the glycoprotein precursor, are the ideal candidates to move forward in clinical trials. However, without a much clearer understanding of correlates of protection in humans, relying on correlates of protection from immunocompromised mice can confound the knowledge needed to develop a successful human vaccine. A recent non-human primate model has demonstrated a spectrum of disease, from asymptomatic to severe, mirroring that observed in humans [98]. However, the severe stages of disease from this model have not been wholly reproducible, as other groups have demonstrated [99] and clinical measures of protection must be established for the model to aid in developing countermeasures, such as vaccines. Additional challenges to vaccine development include strain variance (designing a vaccine that is efficacious across the diverse array of geographic CCHFV clades) [1] and assessing safety profiles for experimental vaccines. Furthermore, the sporadic incidence of human cases, often in rural areas lacking reliable infrastructure for cold chain transport, will impede the return on investment for vaccine manufacturers.

5. Risk Reduction: Animal-Targeted Approaches

5.1. Wild Animals

Controlling human diseases shared with wildlife is complex but can be achieved by different means, including: (i) preventive actions; (ii) arthropod vector control; (iii) host population control through random or selective culling, habitat management, or reproductive control; and (iv) vaccination [100]. Preventive actions (translocation control, barriers, husbandry, etc.) are fundamental approaches for disease control in domestic animals and wildlife but are challenging for CCHFV due to the wide array of host species involved in virus ecology. Notably, the composition of vertebrate hosts on which *Hyalomma* spp. feed is different than for other tick genera. Specifically, immatures feed preferentially on species of the orders Lagomorpha and Rodentia and the class Aves, while adults concentrate mainly on members of the family Bovidae [9]. Thus, efforts to control virus transmission in wildlife will predominantly disrupt transmission by immatures but do not necessarily disrupt virus maintenance by adult *Hyalomma* ticks.

Vector control methods (e.g. acaricides and vaccines) provide viable options for CCHF prevention as discussed later (see Section 6). Population control (random culling, selective culling, habitat modification) may also be considered in endemic areas, as the population density of recognized hosts has been directly associated with the incidence of human disease. For example, *Leporidae* spp. are central in the sylvatic cycle of *Hyalomma* ticks and act as amplifying hosts for immature ticks and CCHFV [9]. Importantly, their population densities have been directly correlated with CCHF incidence. However, population control often consists of an intervention in natural ecosystems and, as such, is often controversial [101]. More socially acceptable alternatives can be expensive and prohibitive based on access to animals; availability of convenient, sensitive, and specific tests; prevalence of infection; and spatial distribution of the target population [100].

Despite presenting its own challenges, wildlife vaccination is another viable option and is proposed by Monath as a Framework III vaccine [102]. A vaccination program targeting vast populations of highly reproductive species over large geographic regions is ambitious. However, tick-borne diseases, including CCHF, are almost always found in geographically circumscribed foci. Therefore, geospatially focused control efforts guided by preceding serosurveillance are a viable option. Some of the barriers to implementing wild animal vaccination include: (i) involvement of multiple species in natural transmission cycles; (ii) safety concerns for non-target species; (iii) high reproductive rates and population turnover; (iv) fastidious feeding behaviors and difficulty in designing effective baits; (v) delivery difficulties due to very high or, conversely, very low population densities of the target species; (vi) environmental concerns and concerns about release of genetically modified organisms; (vii) difficulty in designing an effective formulation for oral immunization; (viii) instability of a vaccine or vector under prevailing environmental conditions; and (ix) requirement for low unit cost and government funding for vaccine purchase and delivery [102]. However, vaccines preventing wildlife diseases have been successfully developed to reduce public health impacts in human populations (e.g., vaccination against rabies of raccoon, coyote, fox, skunk, and bat; against bovine tuberculosis in Eurasian badger), reduce economic effects on the livestock industry (e.g., vaccination against brucellosis in bison and elk; against classical swine fever in wild boar), and address concerns for wildlife conservation (e.g., vaccination against sylvatic plague in black-footed ferret and prairie dog; against white-nose syndrome in hibernating bats) [103]. Oral vaccination is the most common method of wildlife vaccination because it does not require capture and has been proven successful for protecting raccoons, foxes, and coyotes against rabies [104], thereby reducing the risk of rabies exposure in other wild and domestic animals and humans.

The most sustainable approach to eradicating a zoonotic disease from an environment is a reservoir-targeted vaccine (RTV). RTVs have the potential to be effective at breaking the transmission cycle of many tick-borne pathogens including, but not limited to, *Borrelia burgdorferi*, *Borrelia miyamotoi*, *Borrelia mayonii*, *Babesia microti*, and *Anaplasma phagocytophilum* [105]. Multiple studies have shown that free-ranging white-footed mice, the reservoir hosts of Lyme disease, can be effectively vaccinated via RTV bait boxes and that the vaccination both decreases *B. burgdorferi* uptake by larval ticks and decreases transmission from infected ticks to their natural mouse reservoir [105–107]. A similar approach can be envisioned for CCHF, though *Hyalomma* ticks are the long-term reservoir. Decades of effort have shown that targeting non-feeding ticks in the environment either with chemical or physical means does not efficiently reduce tick population density. Therefore, a reservoir-targeted approach focusing on the amplifying hosts of the immature stages of *Hyalomma* ticks, *Leporidae*, would be most successful. A two-pronged approach to RTV that could both interrupt CCHFV transmission and reduce tick populations would likely be most effective.

Non-transmissible vaccines are limited by the quantity and scale of product delivery, whereas transmissible vaccines that can spread from one individual to the next allow much broader coverage with equivalent dosing. A transmissible vaccine must have two crucial properties: A live vectored vaccine must be able to establish and propagate in a given host population, and it must remain attenuated during transmission between vertebrates. A transmissible vaccine with these properties would have numerous benefits that could overcome the challenges of strategies relying on direct vaccination. Ideally, once introduced into a given population, transmissible vaccines would increase herd immunity more efficiently then direct vaccination [108]. However, while transmissible vaccines offer many advantages, serious drawbacks must be addressed in vaccine design or vaccination schemes. These include the potential for reversion to virulence or to inducing clinical signs of disease, short-lived immunity for some pathogens, instability during storage, and contraindications in some subpopulations (e.g., in pregnant animals) [109].

5.2. Domestic Animals

As noted above, livestock, mainly of the family Bovidae, are preferred hosts for adult *Hyalomma* ticks. The livestock–tick and livestock–human interface provide additional opportunities to reduce virus transmission. Operational decisions and livestock preventative health programs are affected by the awareness of risks associated with emergent and endemic diseases of both animals and humans [82]. Although CCHFV infection control may not have perceived animal health benefits, several approaches can reduce virus exposure in animals (and, in turn, humans) and have positive outcomes beyond CCHFV risk reduction. Decreasing the tick burden on animals by physically removing ticks (see Section 4.1) and using acaricides (see Section 6.2) can improve the overall health status of an animal or herd and increase production levels [110,111]. These are significant concerns to herd health, so efforts are made to target breeding programs for tick resistance [112].

Due to limitations in the advancements of a human vaccine against CCHF, the WHO draft roadmap provides alternative vaccination strategies that may be more cost-effective and sustainable at controlling disease, including the development of animal vaccines. Veterinary vaccine development is far less costly; overall, the price of veterinary vaccines, from bench to market, is only ~10% of that of human vaccines. Additional benefits compared to human vaccines include a far less stringent regulatory framework and the ability to directly evaluate candidates in target species. Animal vaccines, however, have their own unique limitations. For example, livestock may require a much larger vaccine dose compared to humans, making certain vaccine platforms (e.g., DNA, RNA, or virus-like particles) extremely cost-prohibitive, and others (e.g., replication-competent platforms or subunit vaccines) are more appropriate for use.

Monath's Framework II vaccines are defined as those that protect humans either indirectly by interrupting transmission of viruses amplified by domesticated animals or directly by preventing virus spread from infected animals to humans, for example, during the slaughter process for CCHFV [102]. The advancement and success of Framework II vaccines for bacterial (*Brucella abortis*, *E. coli* O157) and viral (rabies, influenza, West Nile, Rift Valley fever, and Hendra viruses) pathogens [102] could be capitalized on to develop a multi-valent platform that addresses one or more of these pathogens in conjunction with CCHFV, thus increasing demand.

An important consideration in vaccine design for use in animals is the ability to differentiate previously infected animals from vaccinated animals (DIVA). To accomplish this, the vaccine would need some type of identifiable marker (either a marker fused to the antigen or a deleted portion of the antigen itself) to limit confounding in seroprevalence studies by vaccinated animals. One example of the DIVA principle is in foot-and-mouth disease (FMD) vaccine preparations [113]. These preparations lack non-structural FMD viral proteins. Serological assays detecting nonstructural protein reactivity can therefore be used to indicate previous infection with wild-type virus as opposed to vaccination. CCHFV also encodes non-structural viral glycoproteins, suggesting that a similar approach could be used. However, it is important to note that despite the ideal goal of implementing DIVA practices for transboundary animal diseases, such as CCHFV, designing and evaluating a DIVA vaccine, with the goal of maintaining a high degree of sensitivity, can ultimately increase the total time, cost, and efforts of development. These facets may curtail the benefits offered by a DIVA animal vaccine and are important factors when considering prevention options.

Similar to wildlife (see Section 5.1), RTVs would be a desirable vaccine approach in domestic animals. A two-pronged design conferring anti-tick and anti-CCHFV immunity would be most effective due to simultaneous control of both the virus and the virus reservoir. It is also conceivable that CCHFV vaccines could be added on to conventional common animal vaccines, such as those to prevent bovine herpes virus or lumpy skin disease [114]. In the end, a vaccine will need to be developed that is safe, effective, easy to use, and cost effective. A conceivable approach to a cost-effective strategy could be delivering the immunogen through the feed as shown by using a transgenic tobacco plant expressing CCHFV glycoprotein [115]. As we move to develop safe and effective candidates, we must keep in mind the importance of strategies that aim to increase both vaccine access and demand, all of which are largely dependent on the involvement of donors and international organizations to stimulate and facilitate sustainable adoption [116].

6. Risk Reduction: Tick-Targeted Approaches

6.1. Physical

Manual habitat modification for tick control can affect both immature and adult ticks. For example, the removal of vegetation that shelters immature ticks can reduce their population levels [117]. In addition, physical modification of tick habitats can be used to control adult ticks at the livestock interface. Tick densities decline rapidly in areas where grazing lands are rotated with crops, removing large animal hosts from the accessible range [118] and thus disrupting CCHFV transmission and maintenance by adult ticks. Studies investigating overall tick burden have found that a single spelling/rotation period annually is thought to be promising if combined with acaricide treatments in an integrated pest management approach [119]. However, rotational grazing strategies (frequent movement of cattle between different pens) in the presence [117] and absence [119] of acaricide treatment have also been shown to be effective. Pasture management, including rotational grazing of cattle in Australia and in Zambia, as a tick control strategy is believed to be responsible for an overall decrease in tick burdens on livestock animals [120,121]. Investigating these approaches in areas with a high density of *Hyalomma* ticks would be helpful in determining whether or not they should be promoted in the prevention and control of CCHF.

6.2. Chemical

The use of synthetic acaricides on livestock animals has long been a favored methodology for ectoparasite and tick control throughout the world [122]. Organophosphates are the major chemical acaricides used for ectoparasite control and include compounds, such as pyrethroids, macrocyclic lactones, and amidines, among others [123]. Acaricides are a comparatively inexpensive method for tick control and can be easily administered in multiple ways [124], including dipping, footbaths, or manual sprayers. They vary in efficacy, cost effectiveness, sustainability, and risk to the workers [125]. The ongoing and indiscriminate use of acaricides, however, has led to the development of resistant tick populations, and in recent years, several countries have reported almost complete resistance to most acaricides, presenting a worldwide challenge for ongoing successful tick control [126–130]. Although concerning, resistance in *Hyalomma* ticks has not yet been tested; thus, it is unclear to what extent resistance may contribute to decreased control of *Hyalomma* populations. Additionally, acaricidal compounds can have off-target effects, including toxicity to other organisms, and long-term environmental persistence [131]. For these reasons, the discovery and development of new synthetic acaricidal compounds has not been an actively pursued area of research in recent years [132].

More recently, the focus has been on developing alternative environmentally safe tick control strategies that can circumvent the development of resistance. Historically, many rural farming communities have used plants or plant extracts to control ticks on livestock. The efficacy of some of these ethnobiological methods has been explored in the literature and a recent review. Using in vitro assays, Adenubi et al. (2016, 2018) found that over 200 plant species from several countries around the world demonstrated tick-repellent or acaricidal activity, and some of these also showed marked in vitro acaricidal activity comparable to the efficacy of many synthetic acaricides currently used [133,134]. However, there are several limitations to these studies: (i) a lack of standardized testing methods makes comparisons between studies difficult to interpret and extrapolate for use in controlling ticks on animals; (ii) in vitro data predominate in the literature, and may not adequately recapitulate the efficacy of plants or their extracts in field trials; and (iii) compounds may not persist in the environment due to degradation caused by photo-oxidation, temperature, pH levels, and microbial action [135]. Moreover, differences in cultivating and collecting plant materials (including variations in soil, climate, and other factors) for plant or extract production may affect downstream results [136]. Despite the limitations of current data, based on their historic use by rural livestock farmers, plant-based compounds may be a good future source of effective acaricidal preparations either as an extract or as a source of new synthetic acaricidal compounds [132,134].

6.3. Biological

Due to the limitations of synthetic (and naturally derived) acaricides for tick control, alternative biological methods have been explored, including the use of natural tick predators and pathogenic bacteria, viruses, and fungi. Ants, beetles, and spiders are the major arthropod predators of ticks [137]. Overall, the use of arthropods to reduce tick populations is impractical due to the difficulty in large-scale reproduction and use across broad geographic areas. Birds have been anecdotally documented as capable of preying on ticks, and early evidence from a 1992 study suggested that birds could control ticks carrying Lyme disease in the United States [138]. However, studies have since shown that birds are more likely to play a role in disseminating ticks and their pathogens than in controlling them [51,53,54,139,140]. The systematic breeding and release of birds to control *Hyalomma* tick populations was in fact implemented in Turkey in 2011 and clearly demonstrated the inherent flaws of this strategy. Thousands of helmeted guineafowls (*Numida meleagris*) were introduced to decrease the circulation of CCHFV and its tick vector, but the birds consumed negligible numbers of ticks and instead served as intermediate hosts facilitating the expansion of the *Hyalomma* tick population [60].

Pathogenic bacteria, viruses, and fungi have also been evaluated for their ability to control arthropod populations [141–143]. Fungi showed the most potential and have been effectively used to control medically and agriculturally important arthropods, including German cockroaches and cotton stainer bugs [144–146]. Fungi have been developed industrially with well-established application and dosing methods for commercial use against arthropods [144–146]. *Metarhizium anisopliae* is the most well-studied fungal pathogen of arthropods and is considered to particularly promising due to posing minimal risk to non-target organisms. *M. anisopliae* has been shown to effectively control nymphal *Ixodes* tick [147], cattle tick [148–150], and *Hyalomma* tick populations by inducing high mortality in immature stages and decreasing reproductive fitness in females [151].

Currently, the OIE and the International Research Consortium for Animal Health (STAR-IDAZ) has significant interest in developing anti-tick vaccines to combat a range of tick-borne pathogens. Unlike other tick control measures, the immunological control of ticks is exempt from environmental problems and may provide a broader prevention measure by reducing targeted tick populations in general. Anti-tick vaccines are developed against one of two different types of antigenic targets: "Exposed" antigens that in nature are presented to the animal's immune system (e.g., proteins or peptides secreted in the tick's saliva during attachment and feeding on the host), or "concealed" antigens not normally visible to the host immune mechanisms but associated with a vital function for the tick (e.g., structural components of the tick gut).

To date, efforts to develop anti-tick vaccines have focused on controlling the cattle tick *Rhipicephalus microplus* due to its economic importance. Currently, no anti-tick vaccines based on exposed antigens are commercially available, although some research has been done in this area. Two commercial cattle tick vaccines are available (TickGARD and Gavac), both based on the concealed antigen Bm86, the midgut membrane-bound protein of *R. microplus* [152]. Immunization induces antibodies in the vertebrate host; once ingested, the antibodies interfere with the biological function of Bm86, leading to a reduction in the number, weight, and reproductive capacity of engorged female ticks. Cross-protective efficacy of TickGARD and Gavac has been demonstrated in cattle against *Hyalomma dromedarii* and *Hyalomma anatolicum* ticks (but not *Rhipicephalus appendiculatus* or *Amblyomma variegatum*). The tick antigen subolesin (4D8) was found to be conserved between representatives of six different genera of ticks, including *Hyalomma* [153], offering the prospect for broader cross-species tick vaccines.

Another option for CCHFV control would be vaccinating animals to target the virus in the tick and at the point of entry into the host. Bm86-based vaccines have been shown to induce host antibodies that are ingested when ticks feed on the vaccinated animals and enter the tick gut, thereby causing damage. Thus, it is tempting to speculate that the tick could take up antibodies against CCHFV from appropriately vaccinated animals, potentially resulting in virus neutralization that would prevent virus dissemination to the tick salivary glands or eggs. Such a vaccine may have the potential to reduce infection in difficult to vaccinate hosts (e.g., wild animals), as feeding ticks may be prevented from transmitting the virus.

Experience to date suggests the need to design vaccines tailored to specific local tick populations or to seek universal tick antigens to use as immunogens. Anti-tick vaccines capable of reducing virus transmission will need to be effective against *Hyalomma marginatum*, other *Hyalomma* species, and other competent tick vectors. Existing vaccines and vaccine candidates may also offer some promise. As CCHFV transmission has been demonstrated during co-feeding of infected and naïve ticks on naïve animals [70], decreasing the overall amount of ticks that can acquire virus could play a role in decreasing the spillover events for this disease. However, further research is required to identify potential protective *Hyalomma* antigens together with proof-of-concept studies to demonstrate their efficacy in experimental animals and in the field.

7. Conclusions

The global incidence of CCHF has been rising as changing climatic conditions have led to increased vector survival and expanded disease ranges into previously unaffected geographic regions [4,5]. Additionally, land and habitat fragmentation has been a key factor in exposing animal hosts and humans to tick populations [2]. Although acaricide resistance had not been definitively evaluated in *Hyalomma* ticks, the growing resistance of many tick species to acaricides in endemic countries is also hypothesized to have contributed to higher densities of CCHFV vectors. These factors, compounded by intensified contact between people and livestock, all greatly increase the risk of virus spillover into human populations. The current distribution of the virus across much of Asia, Europe, and Africa is likely to continue to spread over time and poses a significant health risk wherever it may be found. Therapeutic options for treating disease are limited and, in the absence of a human vaccine, alternative approaches to preventing spillover to human populations are essential.

CCHFV spillover events are the summation of environmental, ecological, anthropogenic, and viral genetic factors that must align to result in human disease. The large number of contributors makes predicting future spillover events challenging; however, they also provide opportunities for intervention. We believe that a tiered approach outlining these factors will help to develop a more effective understanding of spillover risk (Figure 2). Factors that include the distribution, density, and infection prevalence in ticks and their animal hosts are the baseline data forming the foundation for targeted risk assessments. Beyond these basic data are human behaviors that predispose certain populations to tick or livestock-associated exposure. Even after virus has breached the physical and intrinsic barriers of the body, disease does not always occur. The virus must be able to replicate effectively (a function of both host and virus genetics) and sufficiently antagonize and evade the immune response. There are numerous gaps in our knowledge of CCHFV in regard to these parameters, necessitating investigations to develop a clear picture of disease risk within defined geographic populations (Table 1).

Given the complexities of CCHFV ecology and endemic circulation as outlined here, we believe that a multisectoral One Health approach that integrates human and animal health, vector biology, and environmental health will be essential in addressing and mitigating the threat of CCHFV in the future. First, building a strong surveillance network of veterinarians, physicians, and epidemiologists can provide important knowledge regarding baseline prevalence of virus in humans, animals, or tick vectors in a given geographic region. Second, as discussed, these data become more powerful when analyzed by modelers and ecologists who can build it into quantitative data-driven risk assessments. Lastly, the data from these risk assessments can be utilized by national One Health networks to target specific tick, animal, or human populations for preemptive interventions that make the most sense for the given population at the given time.

Although not addressed here, the importance of cultivating a national One Health workforce in individual countries cannot be underestimated. In order to adequately integrate data across sectors and implement many of the measures discussed in this paper, a workforce must be developed to carry out and sustain these activities on a national, regional, and individual level. The Tripartite Guide to Addressing Zoonotic Diseases in Countries includes step-by-step recommendations for building the multisectorial One Health network and infrastructure necessary to collect and integrate data, expertise, and personnel from all sectors (environmental, veterinary, medical, etc.) [11]. This process involves activities, such as identifying and convening stakeholders, reviewing current in-country surveillance information, uncovering workforce gaps, developing educational and training programs to address these gaps, and developing a national cohesive strategy for addressing disease.

Ultimately, the way in which data on CCHFV surveillance, risk assessment, and reduction strategies are shared, interpreted, and acted upon must be regionally specific and defined by an interpretation of available data across sectors and the availability of financial resources. While financial resources may be a significant limitation in some countries or sectors, in the long term, many One Health interventions tend to be cost effective and result in improved public health outcomes, decreasing the economic impact that may occur when certain populations are affected by a disease, (e.g., agricultural workers in the case of CCHF). Additionally, coordination across the human, animal, and environmental health sectors can help to reduce costs by avoiding duplication of activities. The economic benefits of improved public health can be used to justify further investment in disease mitigation and aid in supporting a sustained One Health approach to CCHF prevention.

Table 1. Gaps in knowledge of CCHFV and factors contributing to disease. Key considerations and questions regarding parameters of viral maintenance and transmission, including both host and tick factors, that contribute to spillover events (as depicted in Figure 2). These questions aim to highlight areas warranting additional investigation. Reports addressing a subset of these questions currently exist; however, the questions are listed to indicate that more data are required to adequately inform prevention and control efforts.

Parameter	Key Considerations and Questions to Address Gaps in Knowledge of CCHFV
Tick distribution	• Which tick species are competent vectors and what is their vectorial capacity? • What are the molecular determinants of vectorial competency? • Where are these species found and how is their distribution changing? • How do migratory birds, climate change, animal movement, and land fragmentation influence tick range expansion or appearance in new areas? • What is the risk of CCHFV introduction into new geographic areas? • Are there additional cryptic transmission cycles of which we are not yet aware?
Host distribution	• Which hosts are more prominent in virus maintenance? Which are competent at transmitting CCHFV to ticks, and how well do they transmit virus? • Which hosts are amplifiers of key tick species? • What host density is needed to support virus maintenance?
Tick density	• What factors affect immature and adult tick population density? • Can we develop methods to quantitatively assess *Hyalomma* in the field? • How can we predict increases in tick density? • What tick density is needed to support virus maintenance? • What is the relationship between tick density and spillover/outbreak events?
Host density	• What is the relationship between host density and spillover/outbreak events? • Does host diversity and dilution effect play a role for virus maintenance?
Virus prevalence in nature	• How has the virus adapted to support the maintenance cycle in ticks and non-human mammals? • What is the relative prevalence of virus in tick populations? • What is the relative incidence of infection in host species/populations? • What are the best measures of pathogen prevalence? • What are the sensitivity and specificity of serological assays?
Infection intensity *	• Is viremia in host animals necessary to maintain virus in nature or is co-feeding transmission sufficient to maintain the virus in nature? • What is the duration and level of viremia in different host species? • What are transstadial and transovarial transmission rates? • What is viral load and tropism within the tick?
Virus prevalence at slaughter	• What is the prevalence of virus infection in slaughtered animals? • What is the virus load in tissues of infected animals? • How do levels change over time? How stable is the virus in blood and animal tissues?
Biting rate (immatures)	• How relevant is co-feeding transmission? • How efficient is co-feeding transmission and how often does it occur?
Virus survival in blood and tissues of livestock	• How long does infectious virus persist in blood and tissues? • What are the relative levels of infectious virus in tissues? (I.e., handling of which tissues pose the highest risk?)
Tick survival in nature	• How does climate change influence tick life stage survival?
Human behavior	• What are high risk activities/groups? • What is the level of disease awareness in high risk groups? • How widely do individuals in high risk groups accept (or adhere to) preventative measures?
Biting rate (adults)	• What is the affinity of the tick species for biting humans?
Dose and route of exposure	• What is the infectious dose? • What is the rate of subclinical infection? • Does exposure route alter clinical course or outcome? • Are human-to-human transmitted virus strains more virulent than tick-transmitted strains?
Physical and intrinsic barriers (e.g., skin)	• What are the contributions of mucosal and dermal immunity in viral pathogenesis? • What is the stability of virus on skin and mucous membranes? • How does tick saliva influence pathogenesis and virulence? • How quickly is virus transmitted from ticks?
Host tissue susceptibility and permissiveness	• What are underlying risk factors in humans? Genetic factors (e.g., HLAs, polymorphisms in immune response)? • How do pre-existing conditions affect susceptibility to infection/disease? • What is virus tropism in the host and what factors influence it? • What is the relative virulence between virus strains? • What are the viral determinants of virulence? • How does CCHFV suppress the immune response?
Immune response	• What is the contribution of the innate immune response to disease? • Does mitigating the immune response aid or exacerbate disease progression? • What role do innate and adaptive responses play in protection? I.e., what are the correlates of protection?

* Infection intensity refers to the average quantity of virus present in an infected host or tick.

Author Contributions: Conceptualization, T.E.S., S.E.R., J.R.S., D.A.B.; writing—original draft preparation, T.E.S., S.E.R., P.H., M.C., J.R.S., D.A.B.; writing—review and editing, T.E.S., S.E.R., P.H., C.F.S., B.H.B., J.R.S., D.A.B.; visualization, T.E.S., S.E.R., J.R.S., D.A.B.; supervision, L.L.R., C.F.S., B.H.B., J.R.S., D.A.B.; funding acquisition, L.L.R., C.F.S., B.H.B. All authors have read and agreed to the published version of the manuscript.

Funding: This work was supported in part from CDC emerging infectious disease research core funds. Sorvillo and Bird (UC Davis One Health Institute) were supported, in part, by a cooperative agreement from the U.S. Department of Agriculture—Agricultural Research Service (USDA-ARS; 58-8064-9-015). Sergio Rodriguez worked under an appointment to the Centers for Disease Control and Prevention, administered by the Oak Ridge Institute for Science and Education (ORISE) through an interagency agreement between the U.S. Department of Energy (DOE) and the USDA. ORISE is managed by Oak Ridge Associated Universities (ORAU) under contract with DOE.

Acknowledgments: The authors thank Agustín Estrada-Peña, Muhammed Furqan Shahid, Jose de la Fuente, and Sirri Kar for critical review of the manuscript, Stephen Welch for help with figures, and Tatyana Klimova for assistance in editing the manuscript.

Conflicts of Interest: The authors declare no conflict of interest.

Disclaimer: The findings and conclusions in this report are those of the authors and do not necessarily represent the official position of the Centers for Disease Control and Prevention.

References

1. Bente, D.A.; Forrester, N.L.; Watts, D.M.; McAuley, A.J.; Whitehouse, C.A.; Bray, M. Crimean-Congo hemorrhagic fever: History, epidemiology, pathogenesis, clinical syndrome and genetic diversity. *Antivir. Res.* **2013**, *100*, 159–189. [CrossRef] [PubMed]
2. Estrada-Peña, A.; Vatansever, Z.; Gargili, A.; Ergönul, Ö. The trend towards habitat fragmentation is the key factor driving the spread of Crimean-Congo haemorrhagic fever. *Epidemiol. Infect.* **2010**, *138*, 1194–1203. [CrossRef] [PubMed]
3. Spengler, J.R.; Bente, D.A. Crimean-congo hemorrhagic fever in Spain—New arrival or silent resident? *N. Engl. J. Med.* **2017**, *377*. [CrossRef] [PubMed]
4. Ansari, H.; Shahbaz, B.; Izadi, S.; Zeinali, M.; Tabatabaee, S.M.; Mahmoodi, M.; Holakouie-Naieni, K.; Mansournia, M.A. Crimean-Congo hemorrhagic fever and its relationship with climate factors in southeast Iran: A 13-year experience. *J. Infect. Dev. Ctries.* **2014**, *8*, 749–757. [CrossRef] [PubMed]
5. Duygu, F.; Sari, T.; Kaya, T.; Tavsan, O.; Naci, M. The relationship between Crimean-Congo hemorrhagic fever and climate: Does climate affect the number of patients? *Acta Clin. Croat.* **2018**, *57*, 443–448. [CrossRef] [PubMed]
6. Estrada-Peña, A.; Ruiz-Fons, F.; Acevedo, P.; Gortazar, C.; de la Fuente, J. Factors driving the circulation and possible expansion of Crimean-Congo haemorrhagic fever virus in the western Palearctic. *J. Appl. Microbiol.* **2013**, *114*, 278–286. [CrossRef] [PubMed]
7. *Records of Crimean-Congo Hemorrhagic Fever, Communicable Diseases Department*; The Ministry of Health of Turkey: Ankara, Turkey, 2008.
8. Blair, P.W.; Kuhn, J.H.; Pecor, D.B.; Apanaskevich, D.A.; Kortepeter, M.G.; Cardile, A.P.; Ramos, A.P.; Keshtkar-Jahromi, M. An emerging biothreat: Crimean-Congo hemorrhagic fever virus in southern and western Asia. *Am. J. Trop. Med. Hyg.* **2019**, *100*, 16–23. [CrossRef] [PubMed]
9. Spengler, J.R.; Estrada-Peña, A. Host preferences support the prominent role of Hyalomma ticks in the ecology of Crimean-Congo hemorrhagic fever. *PLoS Negl. Trop. Dis.* **2018**, *12*, e0006248. [CrossRef]
10. Zinsstag, J.; Schelling, E.; Waltner-Toews, D.; Tanner, M. From "one medicine" to "one health" and systemic approaches to health and well-being. *Prev. Vet. Med.* **2011**, *101*, 148–156. [CrossRef] [PubMed]
11. FAO; OIE; WHO. *A Tripartite Guide to Addressing Zoonotic Diseases in Countries*; World Health Organization (WHO); Food and Agriculture Organization of the United Nations (FAO); World Organisation for Animal Health (OIE): Geneva, Switzerland, 2019; ISBN 978-92-5-131236-0.
12. Mertens, M.; Schmidt, K.; Ozkul, A.; Groschup, M.H. The impact of Crimean-Congo hemorrhagic fever virus on public health. *Antivir. Res.* **2013**, *98*, 248–260. [CrossRef]
13. Estrada-Peña, A.; Zatansever, Z.; Gargili, A.; Aktas, M.; Uzun, R.; Ergonul, O.; Jongejan, F. Modeling the spatial distribution of Crimean-Congo hemorrhagic fever outbreaks in Turkey. *Vector Borne Zoonotic Dis.* **2007**, *7*, 667–678. [CrossRef] [PubMed]

14. Formenty, P.; Schnepf, G.; Gonzalez-Martin, F.; Bi, Z. International surveillance and control of Crimean-Congo hemorrhagic fever outbreaks. In *Crimean-Congo Hemorrhagic Fever: A Global Perspective*; Springer: Dordrecht, The Netherlands, 2007; pp. 295–303. ISBN 9781402061066.
15. WHO. *WHO CCHF R&D Roadmap Meeting*; World Health Organization (WHO): Geneva, Switzerland, 2018.
16. Spengler, J.R.; Bente, D.A.; Bray, M.; Burt, F.; Hewson, R.; Korukluoglu, G.; Mirazimi, A.; Weber, F.; Papa, A. Second international conference on Crimean-Congo hemorrhagic fever. *Antivir. Res.* **2018**, *150*. [CrossRef] [PubMed]
17. Bodur, H.; Akinci, E.; Ascioglu, S.; Öngürü, P.; Uyar, Y. Subclinical infections with Crimean-Congo hemorrhagic fever virus, Turkey. *Emerg. Infect. Dis.* **2012**. [CrossRef] [PubMed]
18. Gunes, T.; Engin, A.; Poyraz, O.; Elaldi, N.; Kaya, S.; Dokmetas, I.; Bakir, M.; Cinar, Z. Crimean-Congo hemorrhagic fever virus in high-risk population, Turkey. *Emerg. Infect. Dis.* **2009**, *15*, 461–464. [CrossRef] [PubMed]
19. Ertugrul, B.; Kirdar, S.; Ersoy, O.S.; Ture, M.; Erol, N.; Ozturk, B.; Sakarya, S. The seroprevalence of Crimean-Congo haemorrhagic fever among inhabitants living in the endemic regions of Western Anatolia. *Scand. J. Infect. Dis.* **2012**, *44*, 276–281. [CrossRef]
20. Nasirian, H. Crimean-Congo hemorrhagic fever (CCHF) seroprevalence: A systematic review and meta-analysis. *Acta Trop.* **2019**, *196*, 102–120. [CrossRef]
21. Papadopoulos, O.; Koptopoulos, G. Isolation of Crimean-Congo hemorrhagic fever (CCHF) virus from Rhipicephalus bursa ticks in Greece. *Acta Microbiolog. Hell.* **1978**, *23*, 20–28.
22. Papa, A.; Sidira, P.; Larichev, V.; Gavrilova, L.; Kuzmina, K.; Mousavi-Jazi, M.; Mirazimi, A.; Stroher, U.; Nichol, S. Crimean-Congo hemorrhagic fever virus, Greece. *Emerg. Infect. Dis.* **2014**, *20*, 288–290. [CrossRef]
23. Salehi-Vaziri, M.; Baniasadi, V.; Jalali, T.; Mirghiasi, S.M.; Azad-Manjiri, S.; Zarandi, R.; Mohammadi, T.; Khakifirouz, S.; Fazlalipour, M. The first fatal case of Crimean-Congo hemorrhagic fever caused by the AP92-like strain of the Crimean-Congo hemorrhagic fever virus. *Jpn. J. Infect. Dis.* **2016**, *69*, 344–346. [CrossRef]
24. Midilli, K.; Gargili, A.; Ergonul, O.; Elevli, M.; Ergin, S.; Turan, N.; Şengöz, G.; Ozturk, R.; Bakar, M. The first clinical case due to AP92 like strain of Crimean-Congo hemorrhagic fever virus and a field survey. *BMC Infect. Dis.* **2009**, *9*. [CrossRef]
25. Vanhomwegen, J.; Alves, M.J.; Županc, T.A.; Bino, S.; Chinikar, S.; Karlberg, H.; Korukluoğlu, G.; Korva, M.; Mardani, M.; Mirazimi, A.; et al. Diagnostic assays for Crimean-Congo hemorrhagic fever. *Emerg. Infect. Dis.* **2012**, *18*, 1958–1965. [CrossRef] [PubMed]
26. Raabe, V.N. Diagnostic testing for Crimean-Congo hemorrhagic fever. *J. Clin. Microbiol.* **2020**, *58*. [CrossRef] [PubMed]
27. Escadafal, C.; Ölschläger, S.; Avšič-Županc, T.; Papa, A.; Vanhomwegen, J.; Wölfel, R.; Mirazimi, A.; Teichmann, A.; Donoso-Mantke, O.; Niedrig, M. First international external quality assessment of molecular detection of Crimean-Congo hemorrhagic fever virus. *PLoS Negl. Trop. Dis.* **2012**, *6*. [CrossRef] [PubMed]
28. Spengler, J.R.; Bergeron, É.; Rollin, P.E. Seroepidemiological studies of Crimean-Congo hemorrhagic fever virus in domestic and wild animals. *PLoS Negl. Trop. Dis.* **2016**, *10*, e0004210. [CrossRef]
29. Hoogstraal, H. The epidemiology of tick-borne Crimean-Congo hemorrhagic fever in Asia, Europe, and Africa. *J. Med. Entomol.* **1979**, *5*, 307–417. [CrossRef]
30. Spengler, J.R.; Estrada-Peña, A.; Garrison, A.R.; Schmaljohn, C.; Spiropoulou, C.F.; Bergeron, É.; Bente, D.A. A chronological review of experimental infection studies of the role of wild animals and livestock in the maintenance and transmission of Crimean-Congo hemorrhagic fever virus. *Antivir. Res.* **2016**, *135*. [CrossRef]
31. Hawman, D.W.; Meade-White, K.; Haddock, E.; Habib, R.; Scott, D.; Thomas, T.; Rosenke, R.; Feldmann, H. Crimean-Congo hemorrhagic fever mouse model recapitulating human convalescence. *J. Virol.* **2019**, *93*. [CrossRef]
32. Welch, S.R.; Scholte, F.E.M.; Spengler, J.R.; Ritter, J.M.; Coleman-Mccray, J.D.; Harmon, J.R.; Nichol, S.T.; Zaki, S.R.; Spiropoulou, C.F.; Bergeron, E. The Crimean-Congo hemorrhagic fever virus NSm protein is dispensable for growth in vitro and disease in IFNAR$^{-/-}$ mice. *Microorganisms* **2020**, *8*. [CrossRef]
33. Chumakov, M. Contribution to 30 years of investigation of Crimean hemorrhagic fever. *Translyatsia Instituta Polio-Virusnyh Entsefalitov Akademii Medicinskih NauK USSR*, 1974; 22, 5–18. (In Russian) (In English, NAMRU-T950)

34. Causey, O.R.; Kemp, G.E.; Madbouly, M.H.; David-West, T.S. Congo virus from domestic livestock, African hedgehog, and arthropods in Nigeria. *Am. J. Trop. Med. Hyg.* **1970**, *19*, 846–850. [CrossRef]
35. Shanmugam, J.; Smirnova, S.E.; Chumakov, M.P. Presence of antibody to arboviruses of the Crimean haemorrhagic fever-Congo (CHF-Congo) group in human beings and domestic animals in India. *Indian J. Med. Res.* **1976**, *64*, 1403–1413. [PubMed]
36. Goedhals, D.; Paweska, J.T.; Burt, F.J. Long-lived CD8+ T cell responses following Crimean-Congo haemorrhagic fever virus infection. *PLoS Negl. Trop. Dis.* **2017**, *11*. [CrossRef] [PubMed]
37. Burt, F.J.; Samudzi, R.R.; Randall, C.; Pieters, D.; Vermeulen, J.; Knox, C.M. Human defined antigenic region on the nucleoprotein of Crimean-Congo hemorrhagic fever virus identified using truncated proteins and a bioinformatics approach. *J. Virol. Methods* **2013**, *193*, 706–712. [CrossRef] [PubMed]
38. Burt, F.J.; Leman, P.A.; Abbott, J.C.; Swanepoel, R. Serodiagnosis of Crimean-Congo haemorrhagic fever. *Epidemiol. Infect.* **1994**. [CrossRef]
39. Gonzalez, J.-P.; Camicas, J.-L.; Cornet, J.-P.; Wilson, M.L. Biological and clinical responses of West African sheep to Crimean-Congo haemorrhagic fever virus experimental infection. *Res. Virol.* **1998**, *149*, 445–455. [CrossRef]
40. Tignor, G.H.; Smith, A.L.; Casals, J.; Ezeokoli, C.D.; Okoli, J. Close relationship of Crimean hemorrhagic fever-Congo (CHF-C) virus strains by neutralizing antibody assays. *Am. J. Trop. Med. Hyg.* **1980**, *29*, 676–685. [CrossRef]
41. Marriott, A.C.; Ward, V.K.; Higgs, S.; Nuttall, P.A. RNA probes detect nucleotide sequence homology between members of two different nairovirus serogroups. *Virus Res.* **1990**, *16*, 77–81. [CrossRef]
42. Ward, V.K.; Marriott, A.C.; Polyzoni, T.; El-Ghorr, A.A.; Antoniadis, A.; Nuttall, P.A. Expression of the nucleocapsid protein of Dugbe virus and antigenic cross-reactions with other nairoviruses. *Virus Res.* **1992**, *24*, 223–229. [CrossRef]
43. Kondratenko, V.; Blagoveshchenskaya, N.M.; Butenko, A.M.; Vyshnivetskaya, L.K.; Zarubina, L.V.; Milyutin, V.N. Results of virological investigation of ixodid ticks in Crimean hemorrhagic fever focus in Rostov Oblast. *Mater 3 Oblast Nauchn Prakt Konf.* 1970; 3, 29–35. (In Russian). (In English, NAMRU3-T524)
44. Chumakov, M.P. A short story of the investigation of the virus of Crimean hemorrhagic fever. *Translyatsia Instituta Polio-Virusnyh Entsefalitov Akademii Medicinskih NauK USSR*, 1965; 7, 193–196. (In Russian). (In English, NAMRU3-TI89)
45. Pak, T.P.; Daniyarov, O.A.; Kostyukov, M.A.; Bulychev, V.P.; Kuima, A.U. Ecology of Crimean hemorrhagic fever in Tadzhikistan. *Mater Resp. Simp. Kamenyuki Belovezh Pushoha*, 1974; 93–94. (In Russian). (In English, NAMRU3-T968).
46. Swanepoel, R.; Struthers, J.K.; Shepherd, A.J.; McGillivray, G.M.; Nel, M.J.; Jupp, P.G. Crimean-Congo hemorrhagic fever in South Africa. *Am. J. Trop. Med. Hyg.* **1983**, *32*, 1407–1415. [CrossRef]
47. Gargili, A.; Estrada-Peña, A.; Spengler, J.R.; Lukashev, A.; Nuttall, P.A.; Bente, D.A. The role of ticks in the maintenance and transmission of Crimean-Congo hemorrhagic fever virus: A review of published field and laboratory studies. *Antivir. Res.* **2017**, *144*. [CrossRef]
48. Gonzalez, J.P.; Cornet, J.P.; Wilson, M.L.; Camicas, J.L. Crimean-Congo haemorrhagic fever virus replication in adult *Hyalomma truncatum* and *Amblyomma variegatum* ticks. *Res. Virol.* **1991**, *142*, 483–488. [CrossRef]
49. Shepherd, A.J.; Swanepoel, R.; Cornel, A.J.; Mathee, O. Experimental studies on the replication and transmission of Crimean-Congo hemorrhagic fever virus in some African tick species. *Am. J. Trop. Med. Hyg.* **1989**, *40*, 326–331. [CrossRef]
50. Kar, S.; Rodriguez, S.E.; Akyildiz, G.; Cajimat, M.N.B.; Bircan, R.; Mears, M.C.; Bente, D.A.; Keles, A.G. Crimean-Congo hemorrhagic fever virus in tortoises and Hyalomma aegyptium ticks in East Thrace, Turkey: Potential of a cryptic transmission cycle. *Parasites Vectors* **2020**, *13*. [CrossRef] [PubMed]
51. Palomar, A.M.; Portillo, A.; Santibáñez, P.; Mazuelas, D.; Arizaga, J.; Crespo, A.; Gutiérrez, Ó.; Cuadrado, J.F.; Oteo, J.A. Crimean-congo hemorrhagic fever virus in ticks from migratory birds, Morocco. *Emerg. Infect. Dis.* **2013**, *19*, 260–263. [CrossRef] [PubMed]
52. Chisholm, K.; Dueger, E.; Fahmy, N.T.; Samaha, H.A.T.; Zayed, A.; Abdel-Dayem, M.; Villinski, J.T. Crimean-Congo hemorrhagic fever virus in ticks from imported livestock, Egypt. *Emerg. Infect. Dis.* **2012**, *18*, 181–182. [CrossRef]

53. De Liberato, C.; Frontoso, R.; Magliano, A.; Montemaggiori, A.; Autorino, G.L.; Sala, M.; Bosworth, A.; Scicluna, M.T. Monitoring for the possible introduction of Crimean-Congo haemorrhagic fever virus in Italy based on tick sampling on migratory birds and serological survey of sheep flocks. *Prev. Vet. Med.* **2018**, *149*, 47–52. [CrossRef]
54. Buczek, A.M.; Buczek, W.; Buczek, A.; Bartosik, K. The potential role of migratory birds in the rapid spread of ticks and tick-borne pathogens in the changing climatic and environmental conditions in Europe. *Int. J. Environ. Res. Public Health* **2020**, *17*, 2117. [CrossRef]
55. Leggiadro, R.J. Autochthonous Crimean-Congo hemorrhagic fever in Spain. *Pediatr. Infect. Dis. J.* **2017**. [CrossRef]
56. Estrada-Peña, A.; Palomar, A.M.; Santibáñez, P.; Sánchez, N.; Habela, M.A.; Portillo, A.; Romero, L.; Oteo, J.A. Crimean-Congo hemorrhagic fever virus in ticks, Southwestern Europe, 2010. *Emerg. Infect. Dis.* **2012**, *18*, 179–180. [CrossRef] [PubMed]
57. Sureau, P.; Klein, J.M.; Casals, J.; Digoutte, J.; Salaun, J.; Piazak, N.; Calvo, M. Isolation of Thogoto, Wad Medani, Wanowrie and Crimean-Congo haemorrhagic fever viruses from ticks of domestic animals in Iran [*Hyalomma anatolicum, Hyalomma asiaticum, Alveonasus lahorensis*; isolates, epidemiology]. *Ann. Virol.* **1980**, *131*, 185–200.
58. Wilson, M.L.; LeGuenno, B.; Guillaud, M.; Desoutter, D.; Gonzalez, J.P.; Camicas, J.L. Distribution of Crimean-Congo hemorrhagic fever viral antibody in Senegal: Environmental and vectorial correlates. *Am. J. Trop. Med. Hyg.* **1990**, *43*, 557–566. [CrossRef]
59. Logan, T.M.; Linthicum, K.J.; Bailey, C.L.; Watts, D.M.; Moulton, J.R. Experimental transmission of Crimean-Congo hemorrhagic fever virus by *Hyalomma truncatum* koch. *Am. J. Trop. Med. and Hyg.* **1989**, *40*, 207–212. [CrossRef] [PubMed]
60. Şekercioĝlu, Ç.H. Guineafowl, Ticks and Crimean-Congo hemorrhagic fever in Turkey: The perfect storm? *Trends Parasitol.* **2013**, *29*, 1–2. [CrossRef]
61. Plowright, R.K.; Parrish, C.R.; McCallum, H.; Hudson, P.J.; Ko, A.I.; Graham, A.L.; Lloyd-Smith, J.O. Pathways to zoonotic spillover. *Nat. Rev. Microbiol.* **2017**, *15*, 502–510. [CrossRef]
62. Dietz, K. The estimation of the basic reproduction number for infectious diseases. *Stat. Methods Med. Res.* **1993**, *2*, 23–41. [CrossRef]
63. Cooper, B.S. Mathematical modeling of Crimean-Congo hemorrhagic fever transmission. In *Crimean-Congo Hemorrhagic Fever: A Global Perspective*; Springer: Dordrecht, The Netherlands, 2007; pp. 187–203. ISBN 9781402061066.
64. Medley, G.F.; Perry, B.; Young, A.A.S. Preliminary analysis of the transmission dynamics of Theileria parva in eastern Africa. *Parasitology* **1993**, *106*, 251–264. [CrossRef]
65. Ak, Ç.; Ergönül, Ö.; Şencan, İ.; Torunoğlu, M.A.; Gönen, M. Spatiotemporal prediction of infectious diseases using structured Gaussian processes with application to Crimean-Congo hemorrhagic fever. *PLoS Negl. Trop. Dis.* **2018**, *12*. [CrossRef]
66. Norman, R.; Bowers, R.G.; Begon, M.; Hudson, P.J. Persistence of tick-borne virus in the presence of multiple host species: Tick reservoirs and parasite mediated competition. *J. Theor. Biol.* **1999**, *200*, 111–118. [CrossRef] [PubMed]
67. Laurenson, M.K.; Norman, R.A.; Gilbert, L.; Reid, H.W.; Hudson, P.J. Identifying disease reservoirs in complex systems: Mountain hares as reservoirs of ticks and louping-ill virus, pathogens of red grouse. *J. Anim. Ecol.* **2003**, *72*, 177–185. [CrossRef]
68. LoGiudice, K.; Ostfeld, R.S.; Schmidt, K.A.; Keesing, F. The ecology of infectious disease: Effects of host diversity and community composition on Lyme disease risk. *Proc. Natl. Acad. Sci. USA* **2003**, *100*, 567–571. [CrossRef] [PubMed]
69. Jones, L.D.; Davies, C.R.; Steele, G.M.; Nuttall, P.A. A novel mode of arbovirus transmission involving a nonviremic host. *Science* **1987**, *237*, 775–777. [CrossRef] [PubMed]
70. Gordon, S.W.; Linthicum, K.J.; Moulton, J.R. Transmission of Crimean-Congo hemorrhagic fever virus in two species of Hyalomma ticks from infected adults to cofeeding immature forms. *Am. J. Trop. Med. Hyg.* **1993**, *48*, 576–580. [CrossRef]
71. Gonzalez, J.P.; Camicas, J.L.; Cornet, J.P.; Faye, O.; Wilson, M.L. Sexual and transovarian transmission of Crimean-Congo haemorrhagic fever virus in *Hyalomma truncatum* ticks. *Res. Virol.* **1992**, *143*, 23–28. [CrossRef]

72. De la Fuente, J.; Antunes, S.; Bonnet, S.; Cabezas-Cruz, A.; Domingos, A.G.; Estrada-Peña, A.; Johnson, N.; Kocan, K.M.; Mansfield, K.L.; Nijhof, A.M.; et al. Tick-pathogen interactions and vector competence: identification of molecular drivers for tick-borne diseases. *Front. Cell. Infect. Microbiol.* **2017**, *7*. [CrossRef]
73. Ergonul, O.O.; Celikbas, A.; Baykam, N.; Eren, S.; Dokuzoguz, B. Analysis of risk-factors among patients with Crimean-Congo haemorrhagic fever virus infection: Severity criteria revisited. *Clin. Microbiol. Infect.* **2006**. [CrossRef]
74. Mediannikov, O.; Diatta, G.; Fenollar, F.; Sokhna, C.; Trape, J.F.; Raoult, D. Tick-borne rickettsioses, neglected emerging diseases in rural Senegal. *PLoS Negl. Trop. Dis.* **2010**, *4*. [CrossRef]
75. Kamani, J.; Baneth, G.; Apanaskevich, D.A.; Mumcuoglu, K.Y.; Harrus, S. Molecular detection of *Rickettsia aeschlimannii* in Hyalomma spp. ticks from camels (*Camelus dromedarius*) in Nigeria, West Africa. *Med. Vet. Entomol.* **2015**, *29*, 205–209. [CrossRef]
76. Grandi, G.; Chitimia-Dobler, L.; Choklikitumnuey, P.; Strube, C.; Springer, A.; Albihn, A.; Jaenson, T.G.T.; Omazic, A. First records of adult *Hyalomma marginatum* and H. rufipes ticks (Acari: Ixodidae) in Sweden. *Ticks TickBorne Dis.* **2020**, *11*. [CrossRef] [PubMed]
77. Sedaghat, M.; Sarani, M.; Chinikar, S.; Telmadarraiy, Z.; Moghaddam, A.; Azam, K.; Nowotny, N.; Fooks, A.; Shahhosseini, N. Vector prevalence and detection of Crimean-Congo haemorrhagic fever virus in Golestan Province, Iran. *J. Vector Borne Dis.* **2017**, *54*, 353. [CrossRef]
78. Coleman, N.; Coleman, S. Methods of tick removal: A systematic review of the literature. *Australas. Med. J.* **2017**, *10*, 53–62. [CrossRef]
79. Akin Belli, A.; Dervis, E.; Kar, S.; Ergonul, O.; Gargili, A. Revisiting detachment techniques in human-biting ticks. *J. Am. Acad. Dermatol.* **2016**, *75*, 393–397. [CrossRef] [PubMed]
80. Eisen, L. Pathogen transmission in relation to duration of attachment by *Ixodes scapularis* ticks. *Ticks Tick Borne Dis.* **2018**, *9*, 535–542. [CrossRef] [PubMed]
81. Ebel, G.D.; Kramer, L.D. Short report: Duration of tick attachment required for transmission of Powassan virus by deer ticks. *Am. J. Trop. Med. Hyg.* **2004**, *71*. [CrossRef]
82. Booth, T.F.; Steele, G.M.; Marriott, A.C.; Nuttall, P.A. Dissemination, replication, and trans-stadial persistence of Dugbe virus (nairovirus, bunyaviridae) in the tick vector *Amblyomma variegatum*. *Am. J. Trop. Med. Hyg.* **1991**, *45*, 146–157. [CrossRef]
83. Leblebicioglu, H.; Sunbul, M.; Memish, Z.A.; Al-Tawfiq, J.A.; Bodur, H.; Ozkul, A.; Gucukoglu, A.; Chinikar, S.; Hasan, Z. Consensus report: Preventive measures for Crimean-Congo hemorrhagic fever during Eid-al-Adha festival. *Int. J. Infect. Dis.* **2015**, *38*, 9–15. [CrossRef]
84. Mallhi, T.H.; Khan, Y.H.; Sarriff, A.; Khan, A.H. Crimean-Congo haemorrhagic fever virus and Eid-Ul-Adha festival in Pakistan. *Lancet Infect. Dis.* **2016**, *16*, 1332–1333. [CrossRef]
85. Mallhi, T.H.; Khan, Y.H.; Tanveer, N.; Khan, A.H.; Qadir, M.I. Commentary: Surveillance of Crimean-Congo haemorrhagic fever in Pakistan. *Front. Public Health* **2017**, *5*. [CrossRef]
86. Swanepoel, R.; Leman, P.A.; Burt, F.J.; Jardine, J.; Verwoerd, D.J.; Capua, I.; Brückner, G.K.; Burger, W.P. Experimental infection of ostriches with Crimean-Congo haemorrhagic fever virus. *Epidemiol. Infect.* **1998**, *121*, 427–432. [CrossRef]
87. Papa, A.; Papadimitriou, E.; Christova, I. The Bulgarian vaccine Crimean-Congo haemorrhagic fever virus strain. *Scand. J. Infect. Dis.* **2011**, *43*, 225–229. [CrossRef] [PubMed]
88. Tipih, T.; Burt, F.J. Crimean-Congo hemorrhagic fever virus: Advances in vaccine development. *BioRes. Open Access* **2020**, *9*, 137–150. [CrossRef] [PubMed]
89. Dowall, S.D.; Carroll, M.W.; Hewson, R. Development of vaccines against Crimean-Congo haemorrhagic fever virus. *Vaccine* **2017**, *35*, 6015–6023. [CrossRef]
90. Garrison, A.R.; Smith, D.R.; Golden, J.W. Animal models for Crimean-Congo hemorrhagic fever human disease. *Viruses* **2019**, *11*, 590. [CrossRef] [PubMed]
91. Dowall, S.D.; Buttigieg, K.R.; Findlay-Wilson, S.J.D.; Rayner, E.; Pearson, G.; Miloszewska, A.; Graham, V.A.; Carroll, M.W.; Hewson, R. A Crimean-Congo hemorrhagic fever (CCHF) viral vaccine expressing nucleoprotein is immunogenic but fails to confer protection against lethal disease. *Hum. Vaccines Immunother.* **2016**, *12*, 519–527. [CrossRef] [PubMed]
92. Zivcec, M. Characterization of the Interferon αβ Receptor Knockout Mouse Model of Crimean-Congo Hemorrhagic Fever (CCHF) and Assessment of Adenovirus Based CCHF Virus Vaccine Efficacy and Correlates of Protection. Ph.D. Thesis, University of Manitoba, Winnipeg, MB, Canada, 2013.

93. Dowall, S.D.; Graham, V.A.; Rayner, E.; Hunter, L.; Watson, R.; Taylor, I.; Rule, A.; Carroll, M.W. Protective effects of a modified Vaccinia Ankara-based vaccine candidate against Crimean-Congo haemorrhagic fever virus require both cellular and humoral responses. *PLoS ONE* **2016**, *11*. [CrossRef]
94. Kortekaas, J.; Vloet, R.P.M.; Mcauley, A.J.; Shen, X.; Bosch, B.J.; de Vries, L.; Moormann, R.J.M.; Bente, D.A. Crimean-Congo hemorrhagic fever virus subunit vaccines induce high levels of neutralizing antibodies but no protection in STAT1 knockout mice. *Vector Borne Zoonotic Dis.* **2015**, *15*, 759–764. [CrossRef]
95. Bertolotti-Ciarlet, A.; Smith, J.; Strecker, K.; Paragas, J.; Altamura, L.A.; McFalls, J.M.; Frias-Stäheli, N.; García-Sastre, A.; Schmaljohn, C.S.; Doms, R.W. Cellular localization and antigenic characterization of Crimean-Congo hemorrhagic fever virus glycoproteins. *J. Virol.* **2005**, *79*, 6152–6161. [CrossRef]
96. Golden, J.W.; Shoemaker, C.J.; Lindquist, M.E.; Zeng, X.; Daye, S.P.; Williams, J.A.; Liu, J.; Coffin, K.M.; Olschner, S.; Flusin, O.; et al. GP38-targeting monoclonal antibodies protect adult mice against lethal Crimean-Congo hemorrhagic fever virus infection. *Sci. Adv.* **2019**, *5*. [CrossRef]
97. Mousavi-Jazi, M.; Karlberg, H.; Papa, A.; Christova, I.; Mirazimi, A. Healthy individuals' immune response to the Bulgarian Crimean-Congo hemorrhagic fever virus vaccine. *Vaccine* **2012**, *30*, 6225–6229. [CrossRef] [PubMed]
98. Haddock, E.; Feldmann, F.; Hawman, D.W.; Zivcec, M.; Hanley, P.W.; Saturday, G.; Scott, D.P.; Thomas, T.; Korva, M.; Avšič-Županc, T.; et al. A cynomolgus macaque model for Crimean-Congo haemorrhagic fever. *Nat. Microbiol.* **2018**, *3*, 556–562. [CrossRef] [PubMed]
99. Smith, D.R.; Shoemaker, C.J.; Zeng, X.; Garrison, A.R.; Golden, J.W.; Schellhase, C.W.; Pratt, W.; Rossi, F.; Fitzpatrick, C.J.; Shamblin, J.; et al. Persistent Crimean-Congo hemorrhagic fever virus infection in the testes and within granulomas of non-human primates with latent tuberculosis. *PLoS Pathog.* **2019**, *15*. [CrossRef] [PubMed]
100. Gortazar, C.; Diez-Delgado, I.; Barasona, J.A.; Vicente, J.; de La Fuente, J.; Boadella, M. The wild side of disease control at the wildlife-livestock-human interface: A review. *Front. Vet. Sci.* **2014**, *1*, 27. [CrossRef] [PubMed]
101. Artois, M.; Blancou, J.; Dupeyroux, O.; Gilot-Fromont, E. Sustainable control of zoonotic pathogens in wildlife: How to be fair to wild animals? *OIE Rev. Sci. Tech.* **2011**, *30*, 733–743. [CrossRef]
102. Monath, T.P. Vaccines against diseases transmitted from animals to humans: A one health paradigm. *Vaccine* **2013**, *31*, 5321–5338. [CrossRef]
103. Cunningham, A.A.; Daszak, P.; Wood, J.L.N. One health, emerging infectious diseases and wildlife: Two decades of progress? *Philos. Trans. R. Soc. B. Biol. Sci.* **2017**, *372*, 20160167. [CrossRef]
104. Brochier, B.; Blancou, J.; Thomas, I.; Languet, B.; Artois, M.; Kieny, M.-P.; Lecocq, J.-P.; Costy, F.; Desmettre, P.; Chappuis, G.; et al. Use of recombinant vaccinia-rabies glycoprotein virus for oral vaccination of wildlife against rabies: Innocuity to several non-target bait consuming species. *J. Wildl. Dis.* **2013**, *25*, 540–547. [CrossRef]
105. Williams, S.C.; van Oosterwijk, J.G.; Linske, M.A.; Zatechka, S.; Richer, L.M.; Przybyszewski, C.; Wikel, S.K.; Stafford, K.C. Administration of an orally delivered substrate targeting a mammalian zoonotic pathogen reservoir population: Novel application and biomarker analysis. *Vector Borne Zoonotic Dis.* **2020**. [CrossRef]
106. Bhattacharya, D.; Bensaci, M.; Luker, K.E.; Luker, G.; Wisdom, S.; Telford, S.R.; Hu, L.T. Development of a baited oral vaccine for use in reservoir-targeted strategies against Lyme disease. *Vaccine* **2011**, *29*, 7818–7825. [CrossRef]
107. Voordouw, M.J.; Tupper, H.; Önder, Ö.; Devevey, G.; Graves, C.J.; Kemps, B.D.; Brisson, D. Reductions in human Lyme disease risk due to the effects of oral vaccination on tick-to-mouse and mouse-to-tick transmission. *Vector Borne Zoonotic Dis.* **2013**, *13*, 203–214. [CrossRef]
108. Bull, J.J.; Smithson, M.W.; Nuismer, S.L. Transmissible viral vaccines. *Trends Microbiol.* **2018**, *26*, 6–15. [CrossRef] [PubMed]
109. Polak, K.; Kommedal, A.T. (Eds.) *Field Manual for Small Animal Medicine*; Wiley: Hoboken, NJ, USA, 2018; ISBN 9781119243274.
110. Betancur Hurtado, O.J.; Giraldo-Ríos, C. Economic and health impact of the ticks in production animals. In *Ticks Tick-Borne Pathogens*; IntechOpen: London, UK, 2019. [CrossRef]
111. White, N.; Sutherst, R.W.; Hall, N.; Whish-Wilson, P. The vulnerability of the Australian beef industry to impacts of the cattle tick (*Boophilus microplus*) under climate change. *Clim. Change* **2003**, *61*, 157–190. [CrossRef]

112. Hüe, T.; Hurlin, J.C.; Teurlai, M.; Naves, M. Comparison of tick resistance of crossbred Senepol × Limousin to purebred Limousin cattle. *Trop. Anim. Health Prod.* **2014**, *46*, 447–453. [CrossRef] [PubMed]
113. Uttenthal, Å.; Parida, S.; Rasmussen, T.B.; Paton, D.J.; Haas, B.; Dundon, W.G. Strategies for differentiating infection in vaccinated animals (DIVA) for foot-and mouth disease, classical swine fever and avian influenza. *Expert Rev. Vaccines* **2010**, *9*, 73–87. [CrossRef] [PubMed]
114. Farzani, T.A.; Földes, K.; Hanifehnezhad, A.; Ilce, B.Y.; Dagalp, S.B.; Khiabani, N.A.; Ergünay, K.; Alkan, F.; Karaoglu, T.; Bodur, H.; et al. Bovine herpesvirus type 4 (BoHV-4) vector delivering nucleocapsid protein of Crimean-Congo hemorrhagic fever virus induces comparable protective immunity against lethal challenge in IFN$\alpha/\beta\gamma\gamma R^{-/-}$ mice models. *Viruses* **2019**, *11*. [CrossRef]
115. Ghiasi, S.M.; Salmanian, A.H.; Chinikar, S.; Zakeri, S. Mice orally immunized with a transgenic plant expressing the glycoprotein of Crimean-Congo hemorrhagic fever virus. *Clin. Vaccine Immunol.* **2011**, *18*, 2031–2037. [CrossRef]
116. Donadeu, M.; Nwankpa, N.; Abela-Ridder, B.; Dungu, B. Strategies to increase adoption of animal vaccines by smallholder farmers with focus on neglected diseases and marginalized populations. *PLoS Negl. Trop. Dis.* **2019**, *13*, e0006989. [CrossRef]
117. Abdela, N.; Ibrahim, N.; Begna, F. Prevalence, risk factors and vectors identification of bovine anaplasmosis and babesiosis in and around Jimma town, Southwestern Ethiopia. *Acta Tropica* **2018**, *177*, 9–18. [CrossRef]
118. Latif, A.A.; Walker, A.R. An Introduction to the Biology and Control of Ticks in Africa. ICTTD-2 Project. 2004; 1–29.
119. Sutherst, R.W.; Norton, G.A.; Barlow, N.D.; Conway, G.R.; Birley, M.; Comins, H.N. An analysis of management strategies for cattle tick (*Boophilus microplus*) control in Australia. *J. Appl. Ecol.* **1979**, *16*, 359. [CrossRef]
120. Walker, A. *Ticks of Domestic Animals in Africa: A Guide to Identification of Species*; Bioscience Reports: Edinburgh, Scotland, 2003; ISBN 095451730X.
121. Nejash, A.A. Review of economically important cattle tick and its control in Ethiopia. *Vector Biol. J.* **2016**, *1*. [CrossRef]
122. Graf, J.F.; Gogolewski, R.; Leach-Bing, N.; Sabatini, G.A.; Molento, M.B.; Bordin, E.L.; Arantes, G.J. Tick control: An industry point of view. *Parasitology* **2004**, *129*, S427–S442. [CrossRef]
123. Eiden, A.L.; Kaufman, P.E.; Oi, F.M.; Dark, M.J.; Bloomquist, J.R.; Miller, R.J. Determination of metabolic resistance mechanisms in pyrethroid-resistant and fipronil-tolerant brown dog ticks. *Med. Vet. Entomol.* **2017**, *31*, 243–251. [CrossRef] [PubMed]
124. Pavela, R.; Canale, A.; Mehlhorn, H.; Benelli, G. Application of ethnobotanical repellents and acaricides in prevention, control and management of livestock ticks: A review. *Res. Vet. Sci.* **2016**, *109*, 1–9. [CrossRef] [PubMed]
125. De Meneghi, D.; Stachurski, F.; Adakal, H. Experiences in tick control by acaricide in the traditional cattle sector in Zambia and Burkina Faso: Possible environmental and public health implications. *Front. Public Health* **2016**, *4*, 239. [CrossRef]
126. Abbas, R.Z.; Zaman, M.A.; Colwell, D.D.; Gilleard, J.; Iqbal, Z. Acaricide resistance in cattle ticks and approaches to its management: The state of play. *Vet. Parasitol.* **2014**, *203*, 6–20. [CrossRef]
127. Nandi, A.; Sagar, S.V.; Chigure, G.; Fular, A.; Sharma, A.K.; Nagar, G.; Kumar, S.; Saravanan, B.C.; Ghosh, S. Determination and validation of discriminating concentration of ivermectin against *Rhipicephalus microplus*. *Vet. Parasitol.* **2018**, *250*, 30–34. [CrossRef]
128. Pohl, P.C.; Carvalho, D.D.; Daffre, S.; da Silva Vaz, I.; Masuda, A. In vitro establishment of ivermectin-resistant *Rhipicephalus microplus* cell line and the contribution of ABC transporters on the resistance mechanism. *Vet. Parasitol.* **2014**, *204*, 316–322. [CrossRef]
129. Reck, J.; Klafke, G.M.; Webster, A.; Dall'Agnol, B.; Scheffer, R.; Souza, U.A.; Corassini, V.B.; Vargas, R.; dos Santos, J.S.; de Souza Martins, J.R. First report of fluazuron resistance in *Rhipicephalus microplus*: A field tick population resistant to six classes of acaricides. *Vet. Parasitol.* **2014**, *201*, 128–136. [CrossRef]
130. Li, A.Y.; Davey, R.B.; Miller, R.J.; George, J.E. Detection and characterization of amitraz resistance in the southern cattle tick, *Boophilus microplus* (Acari: Ixodidae). *J. Med. Entomol.* **2004**, *41*, 193–200. [CrossRef]
131. Kunz, S.E.; Kemp, D.H. Insecticides and acaricides: Resistance and environmental impact. *Rev. Sci. Tech.* **1994**, *13*, 1249–1286. [CrossRef]

132. Saramago, L.; Gomes, H.; Aguilera, E.; Cerecetto, H.; González, M.; Cabrera, M.; Alzugaray, M.F.; Da Silva Vaz, I., Jr.; Nunes da Fonseca, R.; Aguirre-López, B.; et al. Novel and selective *Rhipicephalus microplus* triosephosphate isomerase inhibitors with acaricidal activity. *Vet. Sci.* **2018**, *5*. [CrossRef] [PubMed]
133. Adenubi, O.T.; McGaw, L.J.; Eloff, J.N.; Naidoo, V. In vitro bioassays used in evaluating plant extracts for tick repellent and acaricidal properties: A critical review. *Vet. Parasitol.* **2018**, *254*, 160–171. [CrossRef] [PubMed]
134. Adenubi, O.T.; Fasina, F.O.; McGaw, L.J.; Eloff, J.N.; Naidoo, V. Plant extracts to control ticks of veterinary and medical importance: A review. *S. Afr. J. Botany* **2016**, *105*, 178–193. [CrossRef]
135. Mulla, M.S.; Su, T. Activity and biological effects of neem products against arthropods of medical and veterinary importance. *J. Am. Mosq. Control Assoc.* **1999**, *15*, 133–152. [PubMed]
136. Heimerdinger, A.; Olivo, C.J.; Molento, M.B.; Agnolin, C.A.; Ziech, M.F.; Scaravelli, L.F.B.; Skonieski, F.R.; Both, J.F.; Charão, P.S. Alcoholic extract of lemongrass (*Cymbopogon citratus*) on the control *of Boophilus microplus* in cattle. *Rev. Bras. Parasitol. Vet.* **2006**, *15*, 37–39.
137. Samish, M.; Alekseev, E. Arthropods as predators of ticks (Ixodoidea). *J. Med. Entomol.* **2001**, *38*, 1–11. [CrossRef]
138. Duffy, D.C.; Downer, R.; Brinkley, C. The effectiveness of Helmeted Guineafowl in the control of the deer tick, the vector of Lyme disease. *Wilson Bull.* **1992**, *104*, 342–345.
139. Hasle, G. Transport of ixodid ticks and tick-borne pathogens by migratory birds. *Front. Cell. Infect. Microbiol.* **2013**, *3*, 48. [CrossRef]
140. Sparagano, O.; George, D.; Giangaspero, A.; Špitalská, E. Arthropods and associated arthropod-borne diseases transmitted by migrating birds. The case of ticks and tick-borne pathogens. *Vet. Parasitol.* **2015**, *213*, 61–66. [CrossRef]
141. Samish, M.; Ginsberg, H.; Glazer, I. Biological control of ticks. *Parasitology* **2004**, *129*, S389–S403. [CrossRef] [PubMed]
142. Assenga, S.P.; You, M.; Shy, C.H.; Yamagishi, J.; Sakaguchi, T.; Zhou, J.; Kibe, M.K.; Xuan, X.; Fujisaki, K. The use of a recombinant baculovirus expressing a chitinase from the hard tick *Haemaphysalis longicornis* and its potential application as a bioacaricide for tick control. *Parasitol. Res.* **2006**, *98*, 111–118. [CrossRef]
143. Fernandes, É.K.K.; Bittencourt, V.R.E.P.; Roberts, D.W. Perspectives on the potential of entomopathogenic fungi in biological control of ticks. *Experimental Parasitol.* **2012**, *130*, 300–305. [CrossRef] [PubMed]
144. Samish, M.; Ginsberg, H.; Glazer, I. Anti-tick biological control agents: Assessment and future perspectives. In *Ticks: Biology, Disease and Control*; Cambridge University Press: Cambridge, UK, 2008; pp. 447–469. ISBN 9780511551802.
145. Santi, L.; e Silva, L.A.D.; da Silva, W.O.B.; Corrêa, A.P.F.; Rangel, D.E.N.; Carlini, C.R.; Schrank, A.; Vainstein, M.H. Virulence of the entomopathogenic fungus Metarhizium anisopliae using soybean oil formulation for control of the cotton stainer bug, *Dysdercus peruvianus*. *World J. Microbiol. Biotechnol.* **2011**, *27*, 2297–2303. [CrossRef]
146. Quesada-Moraga, E.; Santos-Quirós, R.; Valverde-García, P.; Santiago-Álvarez, C. Virulence, horizontal transmission, and sublethal reproductive effects of Metarhizium anisopliae (anamorphic fungi) on the German cockroach (Blattodea: Blattellidae). *J. Invertebr. Pathol.* **2004**, *87*, 51–58. [CrossRef] [PubMed]
147. Stafford, K.C.; Allan, S.A. Field applications of entomopathogenic fungi Beauveria bassiana and Metarhizium anisopliae F52 (Hypocreales: Clavicipitaceae) for the control of *Ixodes scapularis* (Acari: Ixodidae). *J. Med. Entomol.* **2010**, *47*, 1107–1115. [CrossRef]
148. Samish, M.; Gindin, G.; Alekseev, E.; Glazer, I. Pathogenicity of entomopathogenuc fungi to different developmental states of *Rhipicephalus sanguineus* (Acari: Ixodidae). *J. Parasitol.* **2001**, *87*, 1355–1359. [CrossRef]
149. Frazzon, A.P.G.; Vaz Junior, I.D.S.; Masuda, A.; Schrank, A.; Vainstein, M.H. In vitro assessment of Metarhizium anisopliae isolates to control the cattle tick *Boophilus microplus*. *Vet. Parasitol.* **2000**, *94*, 117–125. [CrossRef]
150. Leemon, D.M.; Turner, L.B.; Jonsson, N.N. Pen studies on the control of cattle tick (*Rhipicephalus* (*Boophilus*) *microplus*) with Metarhizium anisopliae (Sorokin). *Vet. Parasitol.* **2008**, *156*, 248–260. [CrossRef]
151. Suleiman, E.A.; Shigidi, M.T.; Hassan, S.M. Metarhizium anisopliae as a biological control agent against *Hyalomma anatolicum* (Acari: Ixodidae). *Pak. J. Biolog. Sci.* **2013**, *16*, 1943–1949. [CrossRef]

152. Rodríguez-Mallon, A. Developing anti-tick vaccines. In *Methods in Molecular Biology*; Humana Press Inc.: New York, NY, USA, 2016; Volume 1404, pp. 243–259.
153. De la Fuente, J.; Almazán, C.; Blas-Machado, U.; Naranjo, V.; Mangold, A.J.; Blouin, E.F.; Gortazar, C.; Kocan, K.M. The tick protective antigen, 4D8, is a conserved protein involved in modulation of tick blood ingestion and reproduction. *Vaccine* **2006**, *24*, 4082–4095. [CrossRef] [PubMed]

Tropical Medicine and Infectious Disease

Review

Rift Valley Fever: Important Considerations for Risk Mitigation and Future Outbreaks

Elysse N. Grossi-Soyster * and A. Desiree LaBeaud *

Department of Pediatrics, Division of Infectious Disease, Stanford University School of Medicine, Stanford, CA 94305, USA

* Correspondence: elysse@stanford.edu (E.N.G.-S.); dlabeaud@stanford.edu (A.D.L.)

Received: 1 May 2020; Accepted: 28 May 2020; Published: 2 June 2020

Abstract: Rift Valley fever virus (RVFV) is a zoonotic phlebovirus of the *Phenuiviridae* family with great opportunity for emergence in previously unaffected regions, despite its current geographical limits. Outbreaks of RVFV often infect humans or domesticated animals, such as livestock, concurrently and occur sporadically, ranging from localized outbreaks in villages to multi-country events that spread rapidly. The true burden of Rift Valley fever (RVF) is not well defined due to underreporting, misdiagnosis caused by the broad spectrum of disease presentation, and minimal access for rapid and accurate laboratory confirmation. Severe symptoms may include hemorrhagic fever, loss of vision, psychological impairment or disturbances, and organ failure. Those living in endemic areas and travelers should be aware of the potential for exposure to ongoing outbreaks or interepidemic transmission, and engage in behaviors to minimize exposure risks, as vaccinations in humans are currently unavailable and animal vaccinations are not used routinely or ubiquitously. The lack of vaccines approved for use in humans is concerning, as RVFV has proven to be highly pathogenic in naïve populations, causing severe disease in a large percent of confirmed cases, which could have considerable impact on human health.

Keywords: Rift Valley fever (RVF); arboviruses; mosquito-borne viruses; zoonoses; One Health; travel medicine; livestock; viral emergence

1. Introduction

Rift Valley fever virus (RVFV) was first isolated in Kenya as a virus with the capacity to infect livestock herds of sheep and cattle, as well as humans [1]. Since its initial discovery, RVFV has been primarily contained within the African continent, with the exception of movement off of the eastern coast of African to the island of Madagascar in 1990 [2,3]. Significant emergence into neighboring regions occurred in the early 2000s when outbreaks were reported in Saudi Arabia and Yemen [4–6]. To this day, much of sub-Saharan Africa and Egypt is endemic for RVFV or has been affected by sporadic outbreaks [7–15].

Transmission of RVFV utilizes mechanisms described by the "One Health" framework, wherein the health and conditions of the environment, animals, and humans intersect and influence each other. Animal transmission is driven by mosquito vectors, primarily *Culex* spp. and floodwater-breeding *Aedes* spp. [9,16–19]. Wild animals have been suspected to contribute to maintenance of RVFV, yet evidence driving such speculation is limited to the presence of antibodies in certain wildlife species [20–22]. Amplification of the virus in mosquitoes [23,24], is linked to mosquito abundance and breeding behaviors that are expanded by periods of heavy rainfall following extreme drought [9,25–31]. Of the many competent vector species [17], infected females of some mosquito species may transmit the virus to their offspring during oviposition, or transovarial transmission (TOT) [32], readily allowing future generations of mosquitoes to transmit RVFV [33]. Transmission in livestock is initiated by mosquito bite and amplified within herds by direct contact with infected bodily fluids, yet there has been little evidence of transmission between animals by way of respiratory droplets and nasal discharge that are characteristic of common respiratory infections [25,34]. There is significant evidence to suggest that vertical transmission may be possible in pregnant animals that are not viremic [35], although findings are limited to laboratory studies and cannot confirm viable offspring following in utero exposure, as infection of pregnant animals typically results in abortion storms that eliminate any viable offspring [36].

Humans can be exposed by mosquito bite or through contact with infected fluids and tissues. Many studies suggest vector-borne transmission is less likely for humans [34]. Zoonotic exposures are driven by many of the occupational and homestead behaviors that are performed with regularity, such as herding, milking, slaughtering animals, and tending to animal health needs in both veterinary and animal health worker capacities [37–40]. Occupational exposures have been shown to elicit a higher incidence than individuals having close contact with or caring for animals at the homestead, and is likely related to contact with a higher volume of animals and their fluids [41]. Aerosolization is also a possible, although unlikely route of transmission, and has been correlated with a higher likelihood of severe disease in laboratory experiments [42].

Despite the presence of RVFV in Africa and the Middle East, emergence of the virus has the potential to cause catastrophic damage to naïve populations of animals and humans. Competent vector species have been identified in many regions that are currently unaffected by RVFV [43–45], providing the ecological support for amplification by mosquito breeding and transovarial transmission (TOT) [32,33]. Rift Valley fever (RVF) causes mild to severe disease in many animal species, with an inverse relationship between the age of the animal and morbidity and mortality, where the younger the animal, the higher the likelihood that the infection will be fatal. Infection in older animals usually produces mild, self-limiting febrile and respiratory symptoms, with a mortality rate ranging from 10% to 30% [46]. Disease severity is also dependent on the species of the animal, and may be specifically virulent in sheep, followed by other commonly domesticated animals such as goats, cattle, buffalo, and camels [45]. While initial symptoms in animals tend to be non-specific, such as diarrhea, vomiting, and respiratory disease, more notable signs of RVFV infection in animals include epistaxis, wasting, spontaneous abortion by pregnant animals, and animal fatalities [25,45].

In humans, RVF disease presentation varies widely, and factors contributing to disease severity are widely unknown. Many experience mild, non-specific, and self-limiting febrile illness that may occasionally present as a biphasic fever with an intermittent remission period of 1–2 days between febrile events [47]. More severe symptoms, typically occurring in up to 8–10% of cases [48], include ocular scarring, central nervous system (CNS) involvement, hemorrhagic fever, organ failure, and death [47,49,50]. RVF can also cause human abortions, still births, and congenital infections [51–53].

Approximately 1–2% of cases experience hemorrhagic fever symptoms, wherein up to 50% of hemorrhagic cases are fatal [10]. The increased risk of fatality with hemorrhagic presentation may be due to a loss of fluids and multisystem shock, organ failure related to loss of blood volume and fluids, or lack of or mismanagement of symptomatic treatment. In vitro studies have suggested that hemorrhage resulting from RVFV infection may be linked to transcription factor IIH (TFIIH) expression levels [54], yet there have yet to be effective treatments for viral hemorrhagic fevers (VHF) beyond basic symptomatic treatment and monitoring [48]. It has been suggested that hemorrhagic cases of RVFV infection may increase the risk of nosocomial transmission for healthcare workers and other individuals providing care [55], yet human-to-human transmission by nosocomial routes of exposure have yet to be documented.

Despite RVF commonly being presented as a mild, self-resolving febrile illness, disease severity has varied by region in epidemiological reports. Publications from Yemen from January 2014 to August 2016 reported hemorrhagic fever in 9% of their anti-RVFV IgM positive hospitalized patients [56], whereas estimates for hemorrhagic symptoms are often limited to 1–2% of cases [57,58]. Early outbreaks in Saudi Arabia experienced approximately double the amount of fatal cases than neighboring Yemen [4], which is likely due to insufficient immunity in the previously naïve community, or increased pathogenicity and disease severity as a result of genomic mutations and reassortments [59]. Variability in disease severity is also seen in neighboring regions, such as countries in East Africa, or intercontinental differences seen in Egypt versus in countries in sub-Saharan Africa and the horn of Africa that are affected by RVFV [7].

Ocular scarring is often reported in 10% of patients [50], while some outbreaks have been associated with more than 40% of patients experiencing loss of vision [60]. Patients experiencing ocular symptoms typically report blurred or loss of vision and posterior eye pain, possibly caused by the development of lesions, edema at the optic disc, or retinal vasculitis or hemorrhaging [49,59,60]. Loss of vision as a result of RVFV infection may be temporary or permanent, depending on the location and severity of the lesions within the ocular tunics. Reports have not distinguished a unilateral or bilateral effect specifically associated with RVFV infection, as confirmed RVFV-positive patients have been documented to suffer retinal scars both unilaterally and bilaterally [49,59].

Multisystem effects of acute RVF are illustrated by involvement of the liver and kidneys, occasionally leading to the onset of hepatitis and nephropathy [34,48,60]. Jaundice and splenomegaly are commonly found in patients during physical exams for diagnosis, and should be monitored carefully to avoid progression to multiple organ failure [48].

Many studies have attempted to identify mechanisms of neurological complications from RVF, yet clear pathways, even those suggesting immune-mediation, have yet to be identified [61–63]. CNS involvement may superficially appear as dizziness or vertigo, confusion and disorientation, and intense headaches, yet may suggest severe underlying manifestations. Meningoencephalopathy can occur in 1–2% of cases and may lead to convulsions, coma, or death [34,60–62,64]. Psychological evaluations of such symptoms suggest CNS involvement may elicit the onset of mental health syndromes, with diagnoses similar to schizophrenia [65–67], and should be taken into consideration when considering immediate treatment and care options. Patients with progression of such syndromes should be evaluated for long term sequelae, as the persistence of psychological syndromes related to RVFV infection has yet to be fully described.

Inconsistent prevalence and incidence of RVFV infection reported is possibly linked to untimely reporting or underreporting of cases or lack of laboratory confirmation in cases of suspected diagnosis [7]. Acute cases are best confirmed by reverse transcriptase polymerase chain reaction (RT-PCR) [68–70], but facilities equipped with the resources, such as skills and instrumentation, required for thorough diagnosis using PCR may be sparse in many endemic regions. Epidemiological studies for assessing RVF burden are often limited to community surveys based on serological analysis of retrospective infection, represented by the presence of immunoglobulin G (IgG) antibodies [68,71]. Detection of immunoglobulin M (IgM) antibodies may be possible [70,71], yet assays designed for IgM detection are notoriously problematic, with potential cross-reactivity and interfering factors (such as rheumatoid factor) leading to inconsistent results, and are therefore not as reliable as PCR diagnostics for acute cases. These analyses may not describe the true burden of RVF in a given population, as acute infections are rarely detected and clinical factors cannot be monitored in real time.

Underreporting may also be due to stigma associated with reporting cases of RVF in animals and humans, which is a phenomenon that is not limited to RVF, but described broadly with infectious diseases throughout history [72,73]. Stigma against RVF survivors has not been reported [38], yet both internal and societal stigmas borne from restrictions with livestock trade and sales may influence downstream behaviors. Trade restrictions for three years are implemented when animal cases are reported and confirmed [74], which may have a major impact on local economies and personal incomes [75]. Additionally, animal infections can trigger a loss of revenue from a reduction in herd size from livestock deaths, and delayed production of sellable animal products due to illness and costly quarantine procedures [75,76]. Spontaneous abortion in pregnant animals also reduces future product generation capacity with the loss of offspring, influencing further financial burden. Community beliefs about processing animal carcasses and use of specific animal parts after death drive personal behaviors that are negligent of the estimated risk of disease exposure, leading individuals to continue engaging in behaviors to avoid bad luck or cultural stigma despite increased risk of personal exposure or continued exposure of other animals, such as skinning animal carcasses before disposal, or harvesting and/or consuming specific organs [77]. Appropriate risk-mitigating behaviors may not be engaged if the perceived risk of exposure is low, or not well understood.

Travel and tourism catalyzes new opportunities for infections in international populations, with the ease of air travel allowing for acutely infected individuals to rapidly reach new destinations [78]. In 2010, Germany had a suspected case of RVF after a tourist visited South Africa, initiating travel warnings for FIFA World Cup events [79,80]. Other regions are suspected to be sources of potential imported cases to Europe, such as Dakar and neighboring regions in Senegal, as they are popular tourist destinations for European residents [81,82]. More recently, an acutely ill individual with persistent symptoms traveled to China after working in Angola in 2016 [83–85]. Traveler-acquired cases of RVF continue to occur, and the public health implications of such cases continue to stress the importance of accurate incidence and prevalence reporting, rapid diagnostic availability and affordability, and the need for a vaccine for human use.

2. Vaccines

RVFV is comprised of a negative-sense, single stranded RNA genome cleaved into three segments varying in size. The functional strength of the genome is driven by the encoding of four structural proteins and two nonstructural proteins. While much is understood about the RVFV genome and viral replication, the genome also contains a sequence for a 78-kDa protein, called Large glycoprotein (LGp) [68,86], which may contribute to viral dissemination in vectors [87,88].

All of the current vaccines licensed for use in animals are generated using viral strains from early outbreaks and isolates, ranging from 1948 to 1977 [68]. Serial passaging has proven some vaccine strains to maintain genomic stability [89], showing minimal risk of reversion to original strain virulence with continual generation and use of the vaccines, but there may be an increased risk of incomplete protection or coverage from newly evolving wildtype strains in circulation. Vaccines designed from early strains do not accommodate the lineage diversity, and may not effectively continue to protect as mutations and reassortments are introduced [58].

Vaccines approved and licensed for use in non-endemic countries are nonexistent, with the exception of MP-12, that has conditional licensure in the US. MP-12 is a live attenuated strain produced by repeat passaging of Egyptian wild type (WT) strain ZH548 in the presence of 5-Fluorouracil, a mutagen that inhibits RNA processing and DNA synthesis [89–91]. Initial viral challenge studies in livestock showed few side effects, with the exception of mild liver necrosis in calves resulting from high dosage administration [92], but many farmers and herders that have used the vaccine have reported low levels of spontaneous abortion when MP-12 is administered to pregnant animals [46,92,93]. Variations on the original MP-12 attenuated vaccination, such as deleting nonstructural genes from the attenuated strain, have also been successful at preventing disease and minimizing lethality, but have yet to be thoroughly tested in the field [94–99]. MP-12 was derived using a human WT strain, and has been proven to be safe for use in humans [100,101], but the duration of neutralizing antibodies has yet to be established, therefore a vaccination schedule has not been proposed. Novel vaccines designed with recombinant proteins, virus-like particles and replicons, and live virus vector-based vaccines which carry DNA-encoding antigens for RVFV in viral vectors such as poxviruses and adenoviruses, have been evaluated for their performance in animal models, such as mice, sheep, and cattle, but have yet to be evaluated for human performance [86,102,103]. Thorough reviews of each of these advanced vaccine designs have been recently summarized [68,86].

At the start of 2019, The Coalition for Epidemic Preparedness Innovations (CEPI) released a call for the development of a human vaccine against RVFV [104], due to RVFV's inclusion on the priority pathogens list released by the World Health Organization [105,106]. CEPI's decision to invest in RVFV was based on the "feasibility of vaccine development and the potential public health impact," [104], which is undoubtedly influenced by the groundwork laid by prior vaccine development for animal use, the increasing frequency of RVFV outbreaks in the last decade, and growing concerns about potential emergence into new territories, such as the United States and the European Union [43,44,68]. Many may question the prioritization of the development of a human vaccine against RVFV, as the mortality rate is reportedly low compared to other, more pressing infections, yet recent viral emergence and extensive outbreaks for which effective vaccines and therapeutics were unavailable, seen with Zika virus (ZIKV) in the Americas and Ebola virus (EBOV) outbreaks in Africa, have shown the need for vaccine development and approval prior to emergent events. The sudden emergence into the Middle East illustrated RVFV's capability to migrate to previously unaffected areas facilitated by human and economic mobility, and the differences in disease severity when previously naïve populations are exposed [4–6,37,56,59,107]. There are currently two RVFV vaccines for humans in Phase II clinical trials [106], with nearly two dozen other vaccine candidates in the preclinical phase of development [106]. If approved, vaccines in humans could be produced for emergency use, to minimize the spread of future human outbreaks within sub-Saharan Africa and the Middle East, or selectively used in individuals with an increased risk of exposure, as the rabies vaccine is given to those with occupational exposure or travel risks. A human RVFV vaccine should also be used to mitigate interepidemic incidence in currently impacted regions. If a safe and effective RVFV vaccine for human use is administered in conjunction with animal vaccines and thorough public health education efforts regarding disease awareness and risk mitigation behaviors, RVFV may be less likely to be considered a priority pathogen in the future.

As novel vaccines are designed and evaluated, it is important to consider the goal of each of the vaccine candidates. Given the complexity of the RVFV transmission cycle, will a vaccine ever effectively prevent RVFV outbreaks in animals and humans, or will the aim of vaccine programs be to minimize disease and fatalities? Further, is it likely that one vaccine will be able to establish immunity in animals and humans, whether in impacted countries or in currently unaffected geographical regions? Future iterations of animal vaccines should be designed for prioritized administration in younger animals, and further vaccines should be evaluated in pregnant animals. RVF-linked abortion has also been reported in humans [51–53], yet currently available vaccines have not been evaluated for their performance in conferring neutralizing antibodies while also avoiding harmful effects on fetal development and viability.

Designs of future vaccine candidates should also consider the economic impact on individuals who will require the vaccine. Vaccines should aim to be economical with minimal burden for cyclical vaccination schedules and boosters. Vaccine workshops in 2011 established that novel vaccines should confer long-duration immunity after a single dosage [102], which has yet to be established with current vaccine iterations that require repeated and seasonal vaccines to induce neutralizing antibodies [94]. Vaccine programs would benefit from vaccines that are designed with the ability to differentiate infected animals from vaccinated animals (DIVA), as many of the current vaccine candidates fail to meet DIVA standards [86]. The ability to effectively differentiate animals that have been naturally infected from those who have been preventatively vaccinated is vital in mounting public health responses to impending and active outbreaks, especially in non-endemic populations. DIVA is also essential to promoting the World Organisation for Animal Health prevention requirements, which precludes any vulnerable animal importation from countries considered to have been infected by RVFV within the last 3 years. Vaccines should also consider the severity and duration of side effects that may impact an animal's ability to produce milks and other products. Farmers, herders, and individuals that maintain livestock herds will be potentially less likely to use a vaccine that produces side effects or physiologically impacts their animals in a way that would impact their production and income.

Vaccination programs aiming to minimize disease severity will require further investigation into the immunopathophysiology of RVFV and identify host immune mechanisms that may increase the likelihood of severe disease symptoms. Severe disease appears to be more often experienced in populations without prior exposure to RVFV, as seen with the emergence into the Arabian Peninsula [4,5,56,59,107], and without further knowledge of causative agents of severe disease, currently naïve populations should be assumed to be at a higher risk for severe morbidity and increased rates of mortality. As witnessed with recent emergence of ZIKV in the Americas [108,109], the impact of unanticipated viral emergence in a novel geographical region or human population can have catastrophic effects.

3. Public Health

Outbreaks continue to occur sporadically, yet their frequency has increased in the last decade, and larger populations are commonly affected. Most recently, an outbreak originating in South Sudan [110] was unable to effectively be contained and may have grown to affect multiple countries across southern and eastern Africa, including Kenya, Rwanda, Uganda, and South Africa [111–114], although some suggest that the outbreaks in each country were not related. Public health measures regarding RVF often focus towards avoiding outbreaks or minimizing the perpetuation of outbreaks rather than minimizing individual exposures and prophylactic measures. Many reports stress the importance of vaccination as the leading way to control outbreaks, yet usage of available vaccines is not standardized or enforced across all endemic regions [7,34,57,74,86]. Additionally, livestock trade between villages and bordering countries may make it difficult to monitor vaccine administration and to achieve effective herd immunity rates to control widespread infection. In addition to aggressive vaccination campaigns, public education is imperative to mitigating continual zoonotic exposures [77].

In naïve regions, concerns regarding use of RVFV as a bioterrorism agent are heightened by the spectrum of disease in animals and humans, and for the potential to devastate large-scale agricultural economies. RVFV is considered a Category A pathogen in the Center for Disease Control and Prevention (CDC)'s Bioterrorism Agent/Disease classifications [115], and as an overlap select agent by the Health and Human Services (HHS) and United States Department of Agriculture (USDA) Federal Select Agents Program [116]. The United States, the Netherlands, the United Kingdom, and many locations in the European Union have been established as potential points for emergence of RVFV in the future due to the availability of mosquito species that are capable of transmitting RVFV, extensive livestock economies and trade, and potential wildlife hosts for interepidemic maintenance [43,44,117,118]. Changes in climate and seasonal extremes may expand the potential for introduction of RVFV to these areas, supporting expansion of vector abundance and conditions for livestock [31]. Air travel contributes to the rapid transport of imported cases, which has been a leading catalyst of the emergence of many arboviral diseases, such as Zika virus and chikungunya virus, that are now established and autochthonously transmitted in previously unaffected areas [57,78,108,109]. While human-to-mosquito transmission is currently speculative for RVFV, many other supportive factors are in place to enhance the likelihood of an emergent outbreak of RVFV.

4. Current Risks and Considerations for Travelers

Travelers should take a specific interest in the risks and exposure opportunities in much of Africa and the Arabian Peninsula. A multifaceted approach to risk mitigation will dramatically reduce the risk of infection, and should include mosquito avoidance, safe practices with animals and animal products, and avoiding contact with infected fluids, tissues, and potential avenues for aerosolization.

Mosquito repellent should be used thoroughly and clothing should be treated to minimize vector biting [119]. *Culex* spp. and *Aedes* spp. that can transmit RVFV are day-biting species, with peaks of feeding near dusk and dawn. Bed nets are an important tool to avoid malaria infection, but mosquito avoidance behaviors should also be practiced during the day for thorough disease avoidance. Mosquito populations in and around the homestead and areas frequently visited can be reduced by dumping out containers that collect standing water, such as tires and buckets, and using air conditioners and screens on windows and doors can dramatically reduce the instance of mosquito exposure. Travelers should also be aware that spraying to control mosquito populations may occur inconsistently, and is often performed in response to an ongoing outbreak, therefore immediate precautions such as the use of personal repellents should be prioritized.

If a traveler anticipates to have contact with animals, specifically domesticated livestock, safe and sanitary measures should be observed in order to reduce contact with potentially infected materials, such as animal bodily fluids and tissues [120,121]. Contact with animals that appear ill or are currently quarantined should be avoided. Consumption of animal products should be limited to items that have been thoroughly cooked to kill any pathogens, including RVFV, as the consumption of meats and milks have been linked to an increased risk of RVFV infection [40,122]. RVFV has the potential to be transmitted by aerosolization [42,49]. Aerosolization is generally a risk observed in occupational settings, such as slaughterhouses, where animal bodily fluids may be aerosolized during processing and handling [41]. Precautions during animal exams, handling animals for milking or other care and maintenance behaviors, and slaughtering or breaking down carcasses should be observed to avoid contact with fluids and blood.

5. Conclusions

RVFV is a complex virus with many possible transmission routes connecting animals and humans, and a wide spectrum of disease without targeted treatment options outside of symptomatic support. RVFV is geographically limited, but many countries contain competent vector species and susceptible hosts that could lead to emergent outbreaks, making RVFV a candidate for an extensive public health burden. Current vaccines are unavailable for humans, which means those living in endemic regions and travelers should practice risk-mitigating behaviors rigorously, and maintain an awareness of the possibility of outbreaks and interepidemic exposure in much of Africa, Saudi Arabia, and Yemen. It is highly encouraged that vaccine development takes into consideration the specific goals of disease prevention, whether future vaccines are utilized to minimize disease or reduce overall incidence.

Author Contributions: Conceptualization, E.N.G.-S. and A.D.L.; investigation, E.N.G.-S.; resources, E.N.G.-S. and A.D.L.; writing—original draft preparation, E.N.G.-S.; writing—review and editing, E.N.G.-S.; supervision, A.D.L. All authors have read and agreed to the published version of the manuscript.

Funding: This research received no external funding.

Conflicts of Interest: The authors declare no conflict of interest.

References

1. Daubney, R.H.J.; Garnham, P.C. Enzootic hepatitis or Rift Valley fever: An undescribed virus disease of sheep, cattle, and man from East Africa. *J. Pathol. Bacteriol.* **1931**, *34*, 545–579. [CrossRef]
2. Morvan, J.; Rollin, P.E.; Laventure, S.; Rakotoarivony, I.; Roux, J. Rift Valley fever epizootic in the central highlands of Madagascar. *Res. Virol.* **1992**, *143*, 407–415. [CrossRef]
3. Morvan, J.; Rollin, P.E.; Roux, J. Rift Valley fever in Madagascar in 1991. Sero-epidemiological studies in cattle. *Rev. Elev. Med. Vet. Pays Trop.* **1992**, *45*, 121–127. [PubMed]
4. Ahmad, K. More deaths from Rift Valley fever in Saudi Arabia and Yemen. *Lancet* **2000**, *356*, 1422. [CrossRef]
5. Centers for Disease Control and Prevention. Update: Outbreak of Rift Valley fever—Saudi Arabia, August–November 2000. *MMWR Morb. Mortal. Wkly. Rep.* **2000**, *49*, 982–985.
6. Centers for Disease Control and Prevention. Outbreak of Rift Valley fever—Saudi Arabia, August–October, 2000. *MMWR Morb. Mortal. Wkly. Rep.* **2000**, *49*, 905–908.
7. Davies, F.G. The historical and recent impact of Rift Valley fever in Africa. *Am. J. Trop. Med. Hyg.* **2010**, *83* (Suppl. 2), 73–74. [CrossRef]
8. Davies, F.G.; Kilelu, E.; Linthicum, K.J.; Pegram, R.G. Patterns of Rift Valley fever activity in Zambia. *Epidemiol. Infect.* **1992**, *108*, 185–191. [CrossRef]
9. Anyamba, A.; Linthicum, K.J.; Small, J.; Britch, S.C.; Pak, E.; de La Rocque, S.; Formenty, P.; Hightower, A.W.; Breiman, R.F.; Chretien, J.P.; et al. Prediction, assessment of the Rift Valley fever activity in East and Southern Africa 2006-2008 and possible vector control strategies. *Am. J. Trop. Med. Hyg.* **2010**, *83* (Suppl. 2), 43–51. [CrossRef]

10. Nguku, P.M.; Sharif, S.K.; Mutonga, D.; Amwayi, S.; Omolo, J.; Mohammed, O.; Farnon, E.C.; Gould, L.H.; Lederman, E.; Rao, C.; et al. An investigation of a major outbreak of Rift Valley fever in Kenya: 2006–2007. *Am. J. Trop. Med. Hyg.* **2010**, *83* (Suppl. 2), 5–13. [CrossRef]
11. Archer, B.N.; Weyer, J.; Paweska, J.; Nkosi, D.; Leman, P.; Tint, K.S.; Blumberg, L. Outbreak of Rift Valley fever affecting veterinarians and farmers in South Africa, 2008. *S. Afr. Med. J.* **2011**, *101*, 263–266. [CrossRef] [PubMed]
12. El Mamy, A.B.; Baba, M.O.; Barry, Y.; Isselmou, K.; Dia, M.L.; El Kory, M.O.; Diop, M.; Lo, M.M.; Thiongane, Y.; Bengoumi, M.; et al. Unexpected Rift Valley fever outbreak, northern Mauritania. *Emerg. Infect. Dis.* **2011**, *17*, 1894–1896. [CrossRef] [PubMed]
13. Imam, I.Z.; Darwish, M.A.; El-Karamany, R. An epidemic of Rift Valley fever in Egypt. 1. Diagnosis of Rift Valley fever in man. *Bull. World Health Organ.* **1979**, *57*, 437–439.
14. Imam, I.Z.; El-Karamany, R.; Darwish, M.A. An epidemic of Rift Valley fever in Egypt. 2. Isolation of the virus from animals. *Bull. World Health Organ.* **1979**, *57*, 441–443.
15. Woods, C.W.; Karpati, A.M.; Grein, T.; McCarthy, N.; Gaturuku, P.; Muchiri, E.; Dunster, L.; Henderson, A.; Khan, A.S.; Swanepoel, R.; et al. An outbreak of Rift Valley fever in Northeastern Kenya, 1997–1998. *Emerg. Infect. Dis.* **2002**, *8*, 138–144. [CrossRef] [PubMed]
16. Ratovonjato, J.; Olive, M.M.; Tantely, L.M.; Andrianaivolambo, L.; Tata, E.; Razainirina, J.; Jeanmaire, E.; Reynes, J.M.; Elissa, N. Detection, isolation, and genetic characterization of Rift Valley fever virus from Anopheles (Anopheles) coustani, Anopheles (Anopheles) squamosus, and Culex (Culex) antennatus of the Haute Matsiatra region, Madagascar. *Vector Borne Zoonotic Dis.* **2011**, *11*, 753–759. [CrossRef] [PubMed]
17. Rostal, M.K.; Evans, A.L.; Sang, R.; Gikundi, S.; Wakhule, L.; Munyua, P.; Macharia, J.; Feikin, D.R.; Breiman, R.F.; Njenga, M.K. Identification of potential vectors of and detection of antibodies against Rift Valley fever virus in livestock during interepizootic periods. *Am. J. Vet. Res.* **2010**, *71*, 522–526. [CrossRef] [PubMed]
18. Sang, R.; Kioko, E.; Lutomiah, J.; Warigia, M.; Ochieng, C.; O'Guinn, M.; Lee, J.S.; Koka, H.; Godsey, M.; Hoel, D.; et al. Rift Valley fever virus epidemic in Kenya, 2006/2007: The entomologic investigations. *Am. J. Trop. Med. Hyg.* **2010**, *83* (Suppl. 2), 28–37. [CrossRef]
19. Turell, M.J.; Lee, J.S.; Richardson, J.H.; Sang, R.C.; Kioko, E.N.; Agawo, M.O.; Pecor, J.; O'Guinn, M.L. Vector competence of Kenyan Culex zombaensis and Culex quinquefasciatus mosquitoes for Rift Valley fever virus. *J. Am. Mosq. Control Assoc.* **2007**, *23*, 378–382. [CrossRef]
20. Davies, F.G. Observations on the epidemiology of Rift Valley fever in Kenya. *J. Hyg.* **1975**, *75*, 219–230. [CrossRef]
21. Evans, A.; Gakuya, F.; Paweska, J.T.; Rostal, M.; Akoolo, L.; Van Vuren, P.J.; Manyibe, T.; Macharia, J.M.; Ksiazek, T.G.; Feikin, D.R.; et al. Prevalence of antibodies against Rift Valley fever virus in Kenyan wildlife. *Epidemiol. Infect.* **2008**, *136*, 1261–1269. [CrossRef] [PubMed]
22. Olive, M.M.; Goodman, S.M.; Reynes, J.M. The role of wild mammals in the maintenance of Rift Valley fever virus. *J. Wildl. Dis.* **2012**, *48*, 241–266. [CrossRef]
23. Sindato, C.; Pfeiffer, D.U.; Karimuribo, E.D.; Mboera, L.E.; Rweyemamu, M.M.; Paweska, J.T. A Spatial Analysis of Rift Valley Fever Virus Seropositivity in Domestic Ruminants in Tanzania. *PLoS ONE* **2015**, *10*, e0131873. [CrossRef] [PubMed]
24. LaBeaud, A.D.; Cross, P.C.; Getz, W.M.; Glinka, A.; King, C.H. Rift Valley fever virus infection in African buffalo (Syncerus caffer) herds in rural South Africa: Evidence of interepidemic transmission. *Am. J. Trop. Med. Hyg.* **2011**, *84*, 641–646. [CrossRef] [PubMed]
25. Gerdes, G.H. Rift Valley fever. *Rev. Sci. Tech.* **2004**, *23*, 613–623. [CrossRef]
26. Sang, R.; Lutomiah, J.; Said, M.; Makio, A.; Koka, H.; Koskei, E.; Nyunja, A.; Owaka, S.; Matoke-Muhia, D.; Bukachi, S.; et al. Effects of Irrigation and Rainfall on the Population Dynamics of Rift Valley Fever and Other Arbovirus Mosquito Vectors in the Epidemic-Prone Tana River County, Kenya. *J. Med. Entomol.* **2017**, *54*, 460–470. [CrossRef]
27. Anyamba, A.; Chretien, J.P.; Small, J.; Tucker, C.J.; Formenty, P.B.; Richardson, J.H.; Britch, S.C.; Schnabel, D.C.; Erickson, R.L.; Linthicum, K.J. Prediction of a Rift Valley fever outbreak. *Proc. Natl. Acad. Sci. USA* **2009**, *106*, 955–959. [CrossRef]
28. Anyamba, A.; Chretien, J.P.; Small, J.; Tucker, C.J.; Linthicum, K.J. Developing global climate anomalies suggest potential disease risks for 2006–2007. *Int. J. Health Geogr.* **2006**, *5*, 60. [CrossRef]

29. Anyamba, A.; Linthicum, K.J.; Small, J.L.; Collins, K.M.; Tucker, C.J.; Pak, E.W.; Britch, S.C.; Eastman, J.R.; Pinzon, J.E.; Russell, K.L. Climate teleconnections and recent patterns of human and animal disease outbreaks. *PLoS Negl. Trop. Dis.* **2012**, *6*, e1465. [CrossRef]
30. Anyamba, A.; Linthicum, K.J.; Tucker, C.J. Climate-disease connections: Rift Valley Fever in Kenya. *Cad. Saude Publica* **2001**, *17*, 133–140. [CrossRef]
31. Anyamba, A.; Small, J.L.; Britch, S.C.; Tucker, C.J.; Pak, E.W.; Reynolds, C.A.; Crutchfield, J.; Linthicum, K.J. Recent weather extremes and impacts on agricultural production and vector-borne disease outbreak patterns. *PLoS ONE* **2014**, *9*, e92538. [CrossRef] [PubMed]
32. Lumley, S.; Horton, D.L.; Hernandez-Triana, L.L.M.; Johnson, N.; Fooks, A.R.; Hewson, R. Rift Valley fever virus: Strategies for maintenance, survival and vertical transmission in mosquitoes. *J. Gen. Virol.* **2017**, *98*, 875–887. [CrossRef] [PubMed]
33. Linthicum, K.J.; Davies, F.G.; Kairo, A.; Bailey, C.L. Rift Valley fever virus (family Bunyaviridae, genus Phlebovirus). Isolations from Diptera collected during an inter-epizootic period in Kenya. *J. Hyg.* **1985**, *95*, 197–209. [CrossRef] [PubMed]
34. World Organization for Animal Health (OIE). Rift Valley Fever (RVF). 2019. Available online: http://www.oie.int/animal-health-in-the-world/animal-diseases/rift-valley-fever/ (accessed on 12 February 2019).
35. Antonis, A.F.; Kortekaas, J.; Kant, J.; Vloet, R.P.; Vogel-Brink, A.; Stockhofe, N.; Moormann, R.J. Vertical transmission of Rift Valley fever virus without detectable maternal viremia. *Vector Borne Zoonotic Dis.* **2013**, *13*, 601–606. [CrossRef]
36. Oymans, J.; Wichgers Schreur, P.J.; van Keulen, L.; Kant, J.; Kortekaas, J. Rift Valley fever virus targets the maternal-foetal interface in ovine and human placentas. *PLoS Negl. Trop. Dis.* **2020**, *14*, e0007898. [CrossRef]
37. Memish, Z.A.; Masri, M.A.; Anderson, J.; Kant, B.D.; Heil, G.L.; Merrill, H.R.; Khan, S.U.; Alsahly, A.; Gray, G.C. Elevated antibodies against Rift Valley fever virus among humans with exposure to ruminants in Saudi Arabia. *Am. J. Trop. Med. Hyg.* **2015**, *92*, 739–743.
38. De St Maurice, A.N.L.; Purpura, L.; Ervin, E.; Tumusiime, A.; Balinandi, S.; Kyondo, J.; Mulei, S.; Tusiime, P.; Manning, C.; Rollin, P.E.; et al. Rift Valley fever: A survey of knowledge, attitudes, and practice of slaughterhouse workers and community members in Kabale District, Uganda. *PLoS Negl. Trop. Dis.* **2018**, *12*, e0006175. [CrossRef]
39. Ng'ang'a, C.M.; Bukachi, S.A.; Bett, B.K. Lay perceptions of risk factors for Rift Valley fever in a pastoral community in northeastern Kenya. *BMC Public Health* **2016**, *16*, 32. [CrossRef]
40. Nicholas, D.E.; Jacobsen, K.H.; Waters, N.M. Risk factors associated with human Rift Valley fever infection: Systematic review and meta-analysis. *Trop. Med. Int. Health* **2014**, *19*, 1420–1429. [CrossRef]
41. Cook, E.A.J.; Grossi-Soyster, E.N.; de Glanville, W.A.; Thomas, L.F.; Kariuki, S.; Bronsvoort, B.M.C.; Wamae, C.N.; LaBeaud, A.D.; Fevre, E.M. The sero-epidemiology of Rift Valley fever in people in the Lake Victoria Basin of western Kenya. *PLoS Negl. Trop. Dis.* **2017**, *11*, e0005731. [CrossRef]
42. Hartman, A.L.; Powell, D.S.; Bethel, L.M.; Caroline, A.L.; Schmid, R.J.; Oury, T.; Reed, D.S. Aerosolized rift valley fever virus causes fatal encephalitis in african green monkeys and common marmosets. *J. Virol.* **2014**, *88*, 2235–2245. [CrossRef] [PubMed]
43. Hartley, D.M.; Rinderknecht, J.L.; Nipp, T.L.; Clarke, N.P.; Snowder, G.D.; National Center for Foreign A, Zoonotic Disease Defense Advisory Group on Rift Valley F. Potential effects of Rift Valley fever in the United States. *Emerg. Infect. Dis.* **2011**, *17*, e1. [CrossRef] [PubMed]
44. Rolin, A.I.; Berrang-Ford, L.; Kulkarni, M.A. The risk of Rift Valley fever virus introduction and establishment in the United States and European Union. *Emerg. Microbes Infect.* **2013**, *2*, e81. [CrossRef]
45. Chevalier, V.; Pepin, M.; Plee, L.; Lancelot, R. Rift Valley fever—A threat for Europe? *Euro Surveill.* **2010**, *15*, 19506. [PubMed]
46. Bird, B.H.; Ksiazek, T.G.; Nichol, S.T.; Maclachlan, N.J. Rift Valley fever virus. *J. Am. Vet. Med. Assoc.* **2009**, *234*, 883–893. [CrossRef] [PubMed]
47. LaBeaud, A.D.; Bashir, F.; King, C.H. Measuring the burden of arboviral diseases: The spectrum of morbidity and mortality from four prevalent infections. *Popul Health Metr.* **2011**, *9*, 1. [CrossRef]
48. Centers for Disease Control and Prevention. Viral Hemorrhagic Fevers (VHFs). 2014; U.S. Department of Health & Human Services. Available online: https://www.cdc.gov/vhf/ (accessed on 1 February 2019).

49. LaBeaud, A.D.; Ochiai, Y.; Peters, C.J.; Muchiri, E.M.; King, C.H. Spectrum of Rift Valley fever virus transmission in Kenya: Insights from three distinct regions. *Am. J. Trop. Med. Hyg.* **2007**, *76*, 795–800. [CrossRef]
50. LaBeaud, A.D.; Pfeil, S.; Muiruri, S.; Dahir, S.; Sutherland, L.J.; Traylor, Z.; Gildengorin, G.; Muchiri, E.M.; Morrill, J.; Peters, C.J.; et al. Factors associated with severe human Rift Valley fever in Sangailu, Garissa County, Kenya. *PLoS Negl. Trop. Dis.* **2015**, *9*, e0003548. [CrossRef]
51. Khan, A.S.S.C. Rift Valley fever: Still an emerging infection after 3500 years. *Lancet Glob. Health* **2016**, *4*, e773–e774. [CrossRef]
52. Baudin, M.; Jumaa, A.M.; Jomma, H.J.E.; Karsany, M.S.; Bucht, G.; Naslund, J.; Ahlm, C.; Evander, M.; Mohamed, N. Association of Rift Valley fever virus infection with miscarriage in Sudanese women: A cross-sectional study. *Lancet Glob. Health* **2016**, *4*, e864–e871. [CrossRef]
53. Adam, I.; Karsany, M.S. Case report: Rift Valley Fever with vertical transmission in a pregnant Sudanese woman. *J. Med. Virol.* **2008**, *80*, 929. [CrossRef] [PubMed]
54. Le May, N.; Dubaele, S.; Proietti De Santis, L.; Billecocq, A.; Bouloy, M.; Egly, J.M. TFIIH transcription factor, a target for the Rift Valley hemorrhagic fever virus. *Cell* **2004**, *116*, 541–550. [CrossRef]
55. Al-Hamdan, N.A.; Panackal, A.A.; Al Bassam, T.H.; Alrabea, A.; Al Hazmi, M.; Al Mazroa, Y.; Al Jefri, M.; Khan, A.S.; Ksiazek, T.G. The Risk of Nosocomial Transmission of Rift Valley Fever. *PLoS Negl. Trop. Dis.* **2015**, *9*, e0004314. [CrossRef] [PubMed]
56. Mansour Shayif, A.A.M.F.; Jamil, S.M.; Saeed, M.A.; Ahmed, M.A. Epidemiological Study on Rift Valley Fever Virus among Humans in Taiz Governorate (Yemen). *Am. J. Clin. Microbiol. Antimicrob.* **2018**, *1*, 6.
57. Hartman, A. Rift Valley Fever. *Clin. Lab. Med.* **2017**, *37*, 285–301. [CrossRef]
58. Baba, M.M.D.; Sang, R.; Villinger, J. Has Rift Valley fever virus evolved with increasing severity in human populations in East Africa? *Emerg. Microbes Infect.* **2016**, *5*, e58–e68. [CrossRef]
59. Al-Hazmi, A.; Al-Rajhi, A.A.; Abboud, E.B.; Ayoola, E.A.; Al-Hazmi, M.; Saadi, R.; Ahmed, N. Ocular complications of Rift Valley fever outbreak in Saudi Arabia. *Ophthalmology* **2005**, *112*, 313–318. [CrossRef]
60. National Institute for Communicable Diseases (NICD). *Healthcare Workers Guidelines on Rift Valley Fever (RVF)*; National Health Laboratory Services (NHLS): Johannesburg, South Africa, 2018.
61. Dodd, K.A.; McElroy, A.K.; Jones, T.L.; Zaki, S.R.; Nichol, S.T.; Spiropoulou, C.F. Rift valley Fever virus encephalitis is associated with an ineffective systemic immune response and activated T cell infiltration into the CNS in an immunocompetent mouse model. *PLoS Negl. Trop. Dis.* **2014**, *8*, e2874. [CrossRef]
62. Walters, A.W.; Kujawa, M.R.; Albe, J.R.; Reed, D.S.; Klimstra, W.B.; Hartman, A.L. Vascular permeability in the brain is a late pathogenic event during Rift Valley fever virus encephalitis in rats. *Virology* **2019**, *526*, 173–179. [CrossRef]
63. Wiley, C.A.; Bhardwaj, N.; Ross, T.M.; Bissel, S.J. Emerging Infections of CNS: Avian Influenza A Virus, Rift Valley Fever Virus and Human Parechovirus. *Brain Pathol.* **2015**, *25*, 634–650. [CrossRef]
64. Terasaki, K.; Makino, S. Interplay between the virus and host in Rift Valley fever pathogenesis. *J. Innate Immun.* **2015**, *7*, 450–458. [CrossRef] [PubMed]
65. Bouloy, M.; Flick, R. Reverse genetics technology for Rift Valley fever virus: Current and future applications for the development of therapeutics and vaccines. *Antivir. Res.* **2009**, *84*, 101–118. [CrossRef] [PubMed]
66. Ludlow, M.; Kortekaas, J.; Herden, C.; Hoffmann, B.; Tappe, D.; Trebst, C.; Griffin, D.E.; Brindle, H.E.; Solomon, T.; Brown, A.S.; et al. Neurotropic virus infections as the cause of immediate and delayed neuropathology. *Acta Neuropathol.* **2016**, *131*, 159–184. [CrossRef] [PubMed]
67. Pinkham, C.; An, S.; Lundberg, L.; Bansal, N.; Benedict, A.; Narayanan, A.; Kehn-Hall, K. The role of signal transducer and activator of transcription 3 in Rift Valley fever virus infection. *Virology* **2016**, *496*, 175–185. [CrossRef] [PubMed]
68. Mansfield, K.L.; Banyard, A.C.; McElhinney, L.; Johnson, N.; Horton, D.L.; Hernandez-Triana, L.M.; Fooks, A.R. Rift Valley fever virus: A review of diagnosis and vaccination, and implications for emergence in Europe. *Vaccine* **2015**, *33*, 5520–5531. [CrossRef]
69. Sall, A.A.; Thonnon, J.; Sene, O.K.; Fall, A.; Ndiaye, M.; Baudez, B.; Mathiot, C.; Bouloy, M. Single-tube and nested reverse transcriptase-polymerase chain reaction for detection of Rift Valley fever virus in human and animal sera. *J. Virol. Methods.* **2001**, *91*, 85–92. [CrossRef]
70. World Health Organization (WHO). Rift Valley Fever. Available online: https://www.who.int/news-room/fact-sheets/detail/rift-valley-fever2018 (accessed on 21 May 2020).

71. Niklasson, B.; Peters, C.J.; Grandien, M.; Wood, O. Detection of human immunoglobulins G and M antibodies to Rift Valley fever virus by enzyme-linked immunosorbent assay. *J. Clin. Microbiol.* **1984**, *19*, 225–229. [CrossRef]
72. Perry, P.; Donini-Lenhoff, F. Stigmatization complicates infectious disease management. *AMA J. Ethics* **2010**, *12*, 225–230.
73. Smith, R.A.; Hughes, D. Infectious Disease Stigmas: Maladaptive in Modern Society. *Commun. Stud.* **2014**, *65*, 132–138. [CrossRef]
74. World Organization for Animal Health (OIE). *Terrestrial Animal Health Code: Recommendations Applicable to OIE Listed Diseases*, 26th ed.; OIE: Paris, France, 2017. Available online: https://www.oie.int/standard-setting/terrestrial-code/access-online/ (accessed on 20 June 2018).
75. Rich, K.M.; Wanyoike, F. An assessment of the regional and national socio-economic impacts of the 2007 Rift Valley fever outbreak in Kenya. *Am. J. Trop. Med. Hyg.* **2010**, *83* (Suppl. 2), 52–57. [CrossRef]
76. Lancelot, R.; Beral, M.; Rakotoharinome, V.M.; Andriamandimby, S.F.; Heraud, J.M.; Coste, C.; Apolloni, A.; Squarzoni-Diaw, C.; de La Rocque, S.; Formenty, P.B.; et al. Drivers of Rift Valley fever epidemics in Madagascar. *Proc. Natl. Acad. Sci. USA* **2017**, *114*, 938–943. [CrossRef] [PubMed]
77. Mutua, E.N.; Bukachi, S.A.; Bett, B.K.; Estambale, B.A.; Nyamongo, I.K. "We do not bury dead livestock like human beings": Community behaviors and risk of Rift Valley fever virus infection in Baringo County, Kenya. *PLoS Negl. Trop. Dis.* **2017**, *11*, e0005582. [CrossRef] [PubMed]
78. Cleton, N.B.; Reusken, C.B.; Wagenaar, J.F.; van der Vaart, E.E.; Reimerink, J.; van der Eijk, A.A.; Koopmans, M.P. Syndromic Approach to Arboviral Diagnostics for Global Travelers as a Basis for Infectious Disease Surveillance. *PLoS Negl. Trop. Dis.* **2015**, *9*, e0004073. [CrossRef] [PubMed]
79. World Health Organization. Rift Valley fever, South Africa-update. *Wkly. Epidemiol. Rec.* **2010**, *85*, 185–186.
80. Boshra, H.; Lorenzo, G.; Busquets, N.; Brun, A. Rift valley fever: Recent insights into pathogenesis and prevention. *J. Virol.* **2011**, *85*, 6098–6105. [CrossRef] [PubMed]
81. Sow, A.F.O.; Ba, Y.; Diallo, D.; Fall, G.; Faye, O.; Bob, N.S.; Loucoubar, C.; Richard, V.; Dia, A.T.; Diallo, M.; et al. Widespread Rift Valley Fever Emergence in Senegal in 2013–2014. *Open Forum Infect. Dis.* **2016**, *3*, ofw149. [CrossRef]
82. Tong, C.; Javelle, E.; Grard, G.; Dia, A.; Lacrosse, C.; Fourie, T.; Gravier, P.; Watier-Grillot, S.; Lancelot, R.; Letourneur, F.; et al. Tracking Rift Valley fever: From Mali to Europe and other countries, 2016. *Euro Surveill.* **2019**, *24*, 1800213. [CrossRef]
83. Fu, X.; Wang, L.; Fang, B.; Ma, R.; Zheng, Y.; Huang, S.; Zhou, P.; Cao, Z.; Tian, J.; Li, S.; et al. Import of Rift Valley fever to China: A potential new threat? *Virol. Sin.* **2016**, *31*, 454–456. [CrossRef]
84. Liu, J.; Sun, Y.; Shi, W.; Tan, S.; Pan, Y.; Cui, S.; Zhang, Q.; Dou, X.; Lv, Y.; Li, X.; et al. The first imported case of Rift Valley fever in China reveals a genetic reassortment of different viral lineages. *Emerg. Microbes Infect.* **2017**, *6*, e4. [CrossRef]
85. Liu, W.; Sun, F.J.; Tong, Y.G.; Zhang, S.Q.; Cao, W.C. Rift Valley fever virus imported into China from Angola. *Lancet Infect. Dis.* **2016**, *16*, 1226. [CrossRef]
86. Faburay, B.; LaBeaud, A.D.; McVey, D.S.; Wilson, W.C.; Richt, J.A. Current Status of Rift Valley Fever Vaccine Development. *Vaccines (Basel)* **2017**, *5*, 29. [CrossRef] [PubMed]
87. Kreher, F.; Tamietti, C.; Gommet, C.; Guillemot, L.; Ermonval, M.; Failloux, A.B.; Panthier, J.J.; Bouloy, M.; Flamand, M. The Rift Valley fever accessory proteins NSm and P78/NSm-GN are distinct determinants of virus propagation in vertebrate and invertebrate hosts. *Emerg. Microbes Infect.* **2014**, *3*, e71. [CrossRef] [PubMed]
88. Weingartl, H.M.; Zhang, S.; Marszal, P.; McGreevy, A.; Burton, L.; Wilson, W.C. Rift Valley fever virus incorporates the 78 kDa glycoprotein into virions matured in mosquito C6/36 cells. *PLoS ONE* **2014**, *9*, e87385. [CrossRef] [PubMed]
89. Lokugamage, N.; Ikegami, T. Genetic stability of Rift Valley fever virus MP-12 vaccine during serial passages in culture cells. *NPJ Vaccines* **2017**, *2*, 1–10. [CrossRef]
90. Wilson, W.C.; Bawa, B.; Drolet, B.S.; Lehiy, C.; Faburay, B.; Jasperson, D.C.; Reister, L.; Gaudreault, N.N.; Carlson, J.; Ma, W.; et al. Evaluation of lamb and calf responses to Rift Valley fever MP-12 vaccination. *Vet. Microbiol.* **2014**, *172*, 44–50. [CrossRef]
91. Caplen, H.; Peters, C.J.; Bishop, D.H. Mutagen-directed attenuation of Rift Valley fever virus as a method for vaccine development. *J. Gen. Virol.* **1985**, *66 Pt 10*, 2271–2277. [CrossRef]

92. Botros, B.; Omar, A.; Elian, K.; Mohamed, G.; Soliman, A.; Salib, A.; Salman, D.; Saad, M.; Earhart, K. Adverse response of non-indigenous cattle of European breeds to live attenuated Smithburn Rift Valley fever vaccine. *J. Med. Virol.* **2006**, *78*, 787–791. [CrossRef]
93. Smith, D.R.; Johnston, S.C.; Piper, A.; Botto, M.; Donnelly, G.; Shamblin, J.; Albarino, C.G.; Hensley, L.E.; Schmaljohn, C.; Nichol, S.T.; et al. Attenuation and efficacy of live-attenuated Rift Valley fever virus vaccine candidates in non-human primates. *PLoS Negl. Trop. Dis.* **2018**, *12*, e0006474. [CrossRef]
94. Labeaud, D. Towards a safe, effective vaccine for Rift Valley fever virus. *Future Virol.* **2010**, *5*, 675–678. [CrossRef]
95. Gowen, B.B.; Westover, J.B.; Sefing, E.J.; Bailey, K.W.; Nishiyama, S.; Wandersee, L.; Scharton, D.; Jung, K.H.; Ikegami, T. MP-12 virus containing the clone 13 deletion in the NSs gene prevents lethal disease when administered after Rift Valley fever virus infection in hamsters. *Front Microbiol.* **2015**, *6*, 651. [CrossRef]
96. Terasaki, K.; Tercero, B.R.; Makino, S. Single-cycle replicable Rift Valley fever virus mutants as safe vaccine candidates. *Virus Res.* **2016**, *216*, 55–65. [CrossRef] [PubMed]
97. Bird, B.H.; Albarino, C.G.; Hartman, A.L.; Erickson, B.R.; Ksiazek, T.G.; Nichol, S.T. Rift valley fever virus lacking the NSs and NSm genes is highly attenuated, confers protective immunity from virulent virus challenge, and allows for differential identification of infected and vaccinated animals. *J. Virol.* **2008**, *82*, 2681–2691. [CrossRef]
98. Ly, H.J.; Lokugamage, N.; Nishiyama, S.; Ikegami, T. Risk analysis of inter-species reassortment through a Rift Valley fever phlebovirus MP-12 vaccine strain. *PLoS ONE* **2017**, *12*, e0185194. [CrossRef] [PubMed]
99. Terasaki, K.; Juelich, T.L.; Smith, J.K.; Kalveram, B.; Perez, D.D.; Freiberg, A.N.; Makino, S. A single-cycle replicable Rift Valley fever phlebovirus vaccine carrying a mutated NSs confers full protection from lethal challenge in mice. *Sci. Rep.* **2018**, *8*, 17097. [CrossRef] [PubMed]
100. Pittman, P.R.; McClain, D.; Quinn, X.; Coonan, K.M.; Mangiafico, J.; Makuch, R.S.; Morrill, J.; Peters, C.J. Safety and immunogenicity of a mutagenized, live attenuated Rift Valley fever vaccine, MP-12, in a Phase 1 dose escalation and route comparison study in humans. *Vaccine* **2016**, *34*, 424–429. [CrossRef] [PubMed]
101. Pittman, P.R.; Norris, S.L.; Brown, E.S.; Ranadive, M.V.; Schibly, B.A.; Bettinger, G.E.; Lokugamage, N.; Korman, L.; Morrill, J.C.; Peters, C.J. Rift Valley fever MP-12 vaccine Phase 2 clinical trial: Safety, immunogenicity, and genetic characterization of virus isolates. *Vaccine* **2016**, *34*, 523–530. [CrossRef]
102. Kortekaas, J.; Zingeser, J.; de Leeuw, P.; de La Rocque, S.; Unger, H.; Moormann, R.J. Rift Valley Fever Vaccine Development, Progress and Constraints. *Emerg. Infect. Dis.* **2011**, *17*, e1. [CrossRef]
103. Ikegami, T. Rift Valley fever vaccines: An overview of the safety and efficacy of the live-attenuated MP-12 vaccine candidate. *Expert Rev. Vaccines* **2017**, *16*, 601–611. [CrossRef]
104. Call for Proposals 3 (CfPi) Human Vaccine Development against Rift Valley Fever and Chikungunya Diseases [Press Release]. Available online: https://cepi.net/wp-content/uploads/2018/12/190104_CfP3_Call-text_v2.0-002.pdf (accessed on 4 January 2019).
105. Coalition for Epidemic Preparedness Innovations (CEPI). Priority Diseases: Rift Valley Fever 2018. Available online: https://cepi.net/research_dev/priority-diseases/ (accessed on 14 March 2019).
106. World Health Organization. 2018 Annual Review of the Blueprint List of Priority Diseases. 2018. Available online: https://www.who.int/blueprint/priority-diseases/en/ (accessed on 14 March 2019).
107. Madani, T.A.; Al-Mazrou, Y.Y.; Al-Jeffri, M.H.; Mishkhas, A.A.; Al-Rabeah, A.M.; Turkistani, A.M.; Al-Sayed, M.O.; Abodahish, A.A.; Khan, A.S.; Ksiazek, T.G.; et al. Rift Valley fever epidemic in Saudi Arabia: Epidemiological, clinical, and laboratory characteristics. *Clin. Infect Dis.* **2003**, *37*, 1084–1092. [CrossRef]
108. Metsky, H.C.; Matranga., C.B.; Wohl, S.; Schaffner, S.F.; Freije, C.A.; Winnicki, S.M.; West, K.; Qu, J.; Baniecki, M.L.; Gladden-Young, A.; et al. Zika virus evolution and spread in the Americas. *Nature* **2017**, *546*, 411–415. [CrossRef]
109. Zhang, Q.; Sun, K.; Chinazzi, M.; Pastore, Y.P.A.; Dean, N.E.; Rojas, D.P.; Merler, S.; Mistry, D.; Poletti, P.; Rossi, L.; et al. Spread of Zika virus in the Americas. *Proc. Natl. Acad. Sci. USA* **2017**, *114*, E4334–E4343. [CrossRef] [PubMed]
110. International Society for Infectious Diseases. Rift Valley Fever—South Sudan. Available online: http://www.promedmail.org/post/20180119.5568536 (accessed on 19 January 2018).
111. International Society for Infectious Diseases. Rift Valley Fever—Eastern Africa: Uganda. Available online: http://www.promedmail.org/post/20180711.5898523 (accessed on 20 June 2018).

112. World Health Organization. Rift Valley Fever—Kenya. 2018. Available online: https://www.who.int/csr/don/18-june-2018-rift-valley-fever-kenya/en/ (accessed on 20 June 2018).
113. Van Vuren, J.K.J.; Patharoo, V.; Ohaebosim, P.; Msimang, V.; Nyokong, B.; Paweska, J.T. Human Cases of Rift Valley fever in South Africa, 2018. *Vector Borne Zoonotic Dis.* **2018**, *18*, 713–715. [CrossRef] [PubMed]
114. International Society for Infectious Diseases. Rift Valley Fever—Eastern Africa: Rwanda. Available online: http://www.promedmail.org/post/20180706.5890383 (accessed on 6 July 2018).
115. Centers for Disease Control and Prevention. Bioterrorism Agents/Diseases. 2018. Available online: https://emergency.cdc.gov/agent/agentlist-category.asp (accessed on 8 February 2019).
116. Federal Select Agent Program. HHS and USDA Select Agents and Toxins. 2018. Available online: https://www.selectagents.gov/selectagentsandtoxinslist.html (accessed on 8 February 2019).
117. Lumley, S.; Hernandez-Triana, L.M.; Horton, D.L.; Fernandez de Marco, M.D.M.; Medlock, J.M.; Hewson, R.; Fooks, A.R.; Johnson, N. Competence of mosquitoes native to the United Kingdom to support replication and transmission of Rift Valley fever virus. *Parasit Vectors.* **2018**, *11*, 308. [CrossRef] [PubMed]
118. Fischer, E.A.; Boender, G.J.; Nodelijk, G.; de Koeijer, A.A.; van Roermund, H.J. The transmission potential of Rift Valley fever virus among livestock in the Netherlands: A modelling study. *Vet. Res.* **2013**, *44*, 58. [CrossRef] [PubMed]
119. Environmental Protection Agency (EPA). Repellents: Protection against Mosquitoes, Ticks, and Other Arthropods. 2018. Available online: https://www.epa.gov/insect-repellents (accessed on 17 October 2018).
120. LaBeaud, A.D.; Kazura, J.W.; King, C.H. Advances in Rift Valley fever research: Insights for disease prevention. *Curr. Opin. Infect. Dis.* **2010**, *23*, 403–408. [CrossRef] [PubMed]
121. Gray, G.C.; Anderson, B.D.; LaBeaud, A.D.; Heraud, J.M.; Fevre, E.M.; Andriamandimby, S.F.; Cook, E.A.; Dahir, S.; de Glanville, W.A.; Heil, G.L.; et al. Seroepidemiological Study of Interepidemic Rift Valley Fever Virus Infection Among Persons with Intense Ruminant Exposure in Madagascar and Kenya. *Am. J. Trop. Med. Hyg.* **2015**, *93*, 1364–1370. [CrossRef] [PubMed]
122. Grossi-Soyster, E.N.; Lee, J.; King, C.H.; LaBeaud, A.D. The influence of raw milk exposures on Rift Valley fever virus transmission. *PLoS Negl. Trop. Dis.* **2019**, *13*, e0007258. [CrossRef]

Tropical Medicine and Infectious Disease

Case Report

Neuroinvasive West Nile Infection with an Unusual Clinical Presentation: A Single-Center Case Series

Nadia Castaldo [1,†], Elena Graziano [1,†], Maddalena Peghin [1,*], Tolinda Gallo [2], Pierlanfranco D'Agaro [3], Assunta Sartor [1], Tiziana Bove [4], Roberto Cocconi [5], Giovanni Merlino [6] and Matteo Bassetti [1,7]

1 Infectious Diseases Division, Department of Medicine, University of Udine and Azienda Sanitaria Universitaria Integrata di Udine, 33100 Udine, Italy; nadiacastaldo.nc@gmail.com (N.C.); elenagraziano22@gmail.com (E.G.); assunta.sartor@asufc.sanita.fvg.it (A.S.); Matteo.Bassetti@hsanmartino.it (M.B.)
2 Department of Prevention, Local Health Unit 4 Medio Friuli, 33100 Udine, Italy; linda.gallo@asufc.sanita.fvg.it
3 Department of Medical, Surgical and Health Sciences, University of Trieste, 34100 Trieste, Italy; dagaro@burlo.trieste.it
4 Anesthesiology and Intensive Care Medicine, Department of Medicine, University of Udine and Azienda Sanitaria Universitaria Integrata di Udine, 33100 Udine, Italy; tiziana.bove@uniud.it
5 SOC Direzione Medica di Presidio, Azienda Sanitaria Universitaria Integrata di Udine, 33100 Udine, Italy; roberto.cocconi@asufc.sanita.fvg.it
6 Clinical Neurology, University of Udine Medical School, 33100 Udine, Italy; giovanni.merlino@asufc.sanita.fvg.it
7 Infectious Diseases Unit, Ospedale Policlinico San Martino-IRCCS, 16132 Genoa, Italy
* Correspondence: maddalena.peghin@gmail.com
† These authors equally contributed to this work.

Received: 16 July 2020; Accepted: 25 August 2020; Published: 31 August 2020

Abstract: The 2018 West Nile Virus (WNV) season in Europe was characterized by an extremely high infection rate and an exceptionally higher burden when compared to previous seasons. Overall, there was a 10.9-fold increase in incidence in Italy, with 577 human cases, 230 WNV neuroinvasive diseases (WNNV) and 42 WNV-attributed deaths. Methods: in this paper we retrospectively reported the neurological presentation of 7 patients admitted to University Hospital of Udine with a diagnosis of WNNV, especially focusing on two patients who presented with atypical severe brain stem involvement. Conclusions: the atypical features of some of these forms highlight the necessity to stay vigilant and suspect the diagnosis when confronted with neurological symptoms. We strongly encourage clinicians to consider WNNV in patients presenting with unexplained neurological symptoms in mild climate-areas at risk.

Keywords: West Nile Virus; mosquitos; vector; Flavivirus; artropodes; neuroinvasiveness

1. Background

West Nile Virus (WNV) is a positive stranded RNA Flavivirus. It has an enzoonotic cycle among birds and mosquitos, and can be rapidly spread to several hosts, including humans. WNV infection in humans might have a wide variety of presentations. In the majority of the cases, it is completely asymptomatic. A minority of the infected subjects develop symptoms, which range from fever, headache, gastrointestinal discomfort, skin rash, to serious neurological disorders.

The 2018 West Nile Virus (WNV) season in Europe was characterized by an earlier start and an exceptionally higher burden when compared to previous seasons [1,2]. Overall, in 2018, 1605 cases of WNV infections were reported in Europe. Italy was the most-affected country, accounting for 39% of the locally acquired infections. Overall, there was a 10.9-fold increase in incidence when compared to 2014–2017 seasons, and a 12-fold increase in comparison to 2017. Infection rates per 100,000 population increased from 0.1 (in 2017) to 1.0 in 2018. Overall, t577 human cases were reported in Italy; WNV neuroinvasive diseases (WNNV) were reported in 230 cases, WNV-attributed deaths were 42. The majority of the infections were from the Veneto, Emilia-Romagna, Lombardia, Piemonte, Sardegna and Friuli Venezia Giulia (FVG) regions [3]. In the FVG region, located in the Northeastern corner of the country, more than 60 WNV infections were reported, including 35 probable and 25 confirmed cases. Among these, nine cases presented as WNNV, and four had a lethal outcome.

The aim of this report is to describe the neurological presentation of seven patients admitted to University Hospital of Udine with a diagnosis of WNNV. We focus on two patients who presented with an atypical manifestation. Table 1 summarizes our clinical experience.

Table 1. West Nile neuroinvasive disease: A single center experience.

Sex, Age (Years)	Previous Clinical History	Date of Infection	Clinical Form	Diagnosis *	Therapy	Outcome and/or Sequelae
Male, 33	Epilepsy	July 2018	VII cranial nerve palsy	Serology (IgM) in serum and LCR	Steroid	Healed
Female, 15	Previously healthy	July 2018	Fever, cephalea, Parkinsonism, left arm strength deficit, tongue fasciculation	Serology (IgM and IgG) in CSF	None	Healed
Male, 87	Ischemic cardiomyopathy	August 2018	Fever, meningismus and neuropathic pain, limbs waxy flexibility and catatonic states.	rt-PCR in both urine and CSF; serology (IgM and IgG) in both CSF and serum	IVIG and steroids	Residual disability, depression
Male, 56	Renal transplant, previous CMV and VZV reactivation	August 2018	Fever, confusion, meningismus	Serology (IgM) in serum; rt-PCR in urine	IVIG and steroids	Healed
Male, 57	Autoimmune glomerulonephritis in immunosuppressive treatment	September 2018	Rhombencephalitis, coma, intracranial hemorrhage	Serology (IgM) in serum; rt-PCR in CSF, urine and blood	IVIG and steroids	Death
Female, 78	Diabetes, hypertension, non-Hodgkin lymphoma	October 2018	Confusion, stupor, coma	rt-PCR in CSF, urine and blood		Death
Female, 77	B chronic lymphocytic leukemia, hypogammaglobulinaemia, hypertension, diabetes, previous CMV reactivation	November 2018	Fever, confusion, 4 limbs hyposthenia, acute kidney injury, coma	rt-PCR in CSF and urine	IVIG and steroids	Partial improvement, residual disability

Abbreviations—rt-PCR: reverse-transcription polymerase chain reaction, IgM: immunoglobulin M, IgG; immunoglobulin G; IVIG: intravenous immunoglobulin; CSF: cerebrospinal fluid; CMV: cytomegalovirus; VZV: varicella zoster virus. * Only positive results are reported.

2. Case 1

An 87-year-old man was admitted to Cardiology Intensive Care Unit on August 2018 for non-ST-segment-elevation myocardial infarction. His prior history was remarkable for ischemic cardiomyopathy and arterial hypertension. No invasive procedure was performed during the hospitalization. On day 3, the patient presented abrupt onset of fever, meningeal syndrome, generalized stiffness and diffuse neuropathic pain. He was alert and oriented and showed no focal deficit. His blood exams showed leucopenia (3580/mmc white blood cells (WBC)), thrombocytopenia (47,000 platelets/μL) and elevation of liver enzymes (aspartate aminotransferase (AST) 95 UI/L, alanine aminotransferase (ALT) 80 UI/L). The cerebrospinal fluid (CSF) was clear and showed hypoglycorrhachia (25%), hyperproteinorrachia (637 g/L) and hypercellularity (85 cells/μL, 48% polymorphonuclear). Gram stain, culture and serologic tests for common neurotropic viruses were negative. Brain magnetic resonance imaging (MRI) showed no abnormalities. Laboratory diagnosis of WNV infection was accomplished through biologic fluids testing for WNV- specific IgG and IgM antibodies, and reverse-transcription polymerase chain reaction (rt-PCR). Patient n.1 was tested positive for WNV rt-PCR on serum and blood, and specific IgM and IgG were found in both CSF and serum. A 5-day course of intravenous immunoglobulin (IVIG) at 0.4 g/kg/day and high-dose steroid therapy (500 mg/day for 2 days followed by 1 mg/kg/day with progressive de-escalation) were administered. The patient's general status slowly improved. On day 10, the patient experienced an abrupt neurologic deterioration; he was found unconscious, with a Glasgow Coma Scale (GCS) of 3. Any organic cause of encephalopathy was excluded; his blood tests, head computed tomography (CT) scan, electrocardiogram (ECG) and electroencephalography (EEG) were inconclusive. CSF was checked again and showed residual cellularity (24 cells/mmc, 60% lymphocytes, 30% polymorphs). The patient was intubated and admitted to the intensive care unit (ICU). A few hours later, he spontaneously recovered and was extubated. During the next 2 weeks of hospitalization, the patient manifested three other similar episodes of "intermittent coma". A conservative management was performed. He experienced a partial recovery and was discharged.

3. Case 2

A 57-year-old man with an autoimmune glomerulonephritis (being treated with steroids and mycophenolate) presented to the emergency department, with a history of three days of fever, confusion, diplopia, opsoclonus, multifocal myoclonus and generalized tremor. Blood exams and CT scan were unremarkable. Lumbar puncture revealed normal opening pressure, and the CSF analysis identified 12 cells/μL (30% polymorphonuclear, 60% lymphocytes), protein count of 200 g/L and glucose count of 70 mg/dL. The EEG showed slow bilateral diffuse slow waves. A presumptive diagnosis of rhombencephalitis with opsoclonus-myoclonus syndrome was made, and an empirical therapy with ampicillin, ceftriaxone, acyclovir and dexamethasone was started. Contrast enhanced brain magnetic resonance imaging (MRI) revealed no acute process or abnormality. During the hospitalizations, the patient's condition worsened. He progressively became comatose (GCS of 4) and required ICU admission and mechanical ventilation. The laboratory documented the presence of WNV-RNA in CSF, urine and blood samples, and WNV-specific immunoglobulin M (IgM) in blood. Therefore, high doses of IVIGs and steroids were administered. On day 4, fixed bilateral mydriasis appeared. A CT scan showed massive intraparenchymal hemorrhage associated with fourth ventricle compression and tonsillar herniation. The following day, the patient expired. Post-mortem macroscopic examination of the brain showed diffuse malacia.

4. Discussion

WNV transmission season 2018 has been characterized by an extremely high infection rate. Overall, 2083 autochthonous infections were reported in 2018 in Europe, with an increase of 7.2-fold when compared to 2017 [1,2].

WNV diagnosis is mainly formulated by detecting specific IgM and IgG antibodies directed towards WNV, or by testing for viral nucleic acid in CSF, tissue, blood, or other body fluid. Evidence of IgM in CSF is the cornerstone of the WNV diagnosis because IgM does not cross the blood brain barrier. Nevertheless, serology has some pitfalls: IgM remains detectable for 2 to 12 months. Furthermore, the usual serological enzyme-linked immunosorbent assay (ELISA) may cross-react with other flavivirus infections or vaccination. There is a great variety of serological methods, but none of these provide a definitive diagnosis.

Nevertheless, any positive result requires a confirmation with more specific tests. Thus, when IgM are found positive, seroconversion (by subsequent convalescent sample testing, hemagglutination inhibition, IgG ELISA, or plaque reduction neutralization assay) should be demonstrated [4].

However, IgM and IgG testing might be insufficient in some cases. Antibodies might not be present during the window period, as to cause a delay in the diagnosis. Furthermore, special categories, such as immunocompromised patients, might not be able to mount an adequate serological response. In this setting, molecular testing for WNV could be useful; however it is known that viremia is generally rapidly vanishing, and lasts between 3 and 14 days. Prolonged viremia has been reported in immuno-compromised patients [5].

Several studies demonstrated that WNV RNA determination in whole blood shows higher reliability compared to other samples [6,7]. Detection of WNV in urine is not included in the current case definition of WNV disease postulated by the European Centers for Disease Control. Although WNV viral load in CSF and blood is usually transient, shedding in the urine can continue for longer periods after the acute phase [8]. Indeed, the US Centers for Disease Control guidelines have included the detection of RNA in body fluids other than blood and LCR, including viruria, in the diagnostic criteria for WNV diagnosis confirmation [9].

Furthermore, pathology may reveal peculiar histological pictures. Autopsy reports have shown perivascular lymphoplasmacytic infiltrates, both in leptomeninges and parenchyma, microglial nodules or acute neuronal necrosis. Inflammation generally involves deep nuclei, the cerebellum and brainstem, with neuronal loss and neuronophagia. In our cases, WNV was present in brain matter, a finding that suggests a strong correlation between its neurotropism with the distribution of the cerebral lesions [10].

Most WNV infections are usually asymptomatic; 30% of cases present with a febrile illness characterized by mild symptoms that typically last up to a week. Less than 1% of infections present with central nervous system involvement [11]. We reported two WNV cases consistent with severe brain stem involvement. Although there was no electrophysiological or neuroradiology evidence, both patients presented symptoms due to brainstem and basal ganglia damage. Case 1 showed neuropsychiatric disorder and loss of consciousness with no apparent organic explanation. Case 2 presented unique findings consistent with opsoclonus-myoclonus syndrome (OPS) complicated with a massive hemorrhagic stroke. To our knowledge, 14 cases of WNNV-related OPS have been reported to date [12,13]. The majority of the cases were reported among patients affected by neoplasm or immune system deficit, as well as our case. Confusion and delirium, which were the main presenting symptoms in patient n.2, were reported in 30% of the cases, usually after hospital admission. Overall, no specific MRI findings were described in WNV associated-OPS. However, brainstem infectious involvement has been speculated to cause secondary cortical hypoperfusion and subsequent dysfunction [14].

Overall, WNV-associated CNS vascular events are rare [12,15,16]. The main hypothesis regarding the pathophysiology of these events is an inflammatory vasculopathy secondary to WNV CNS invasion. It has been suggested that WNV reaches CNS through two different routes: spreading within the axons of peripheral nerves to CNS, and via the blood system. A number of in vitro analyses demonstrated that, similarly to other flaviviruses, WNV can directly infect the endothelial cells. Once penetrated in the CNS, WNV elicit complex immunitary responses, which involve both intrinsic CNS immune cells (microglia and astrocytes) and extrinsic defenses (T cells, B cells, and monocytes/macrophages), thereby, leading to cytokines release, perivascular inflammation, and neuronal proliferation [17–19]. This could provide an explanation of the WNV-related vasculitis.

Treatment for WNV infection is mainly supportive. No therapy has been proven useful in a randomized controlled trial. Few reports and preclinical studies suggest the utility of immunomodulation through IVIG, plasmapheresis, WNV-specific neutralizing antibodies, corticosteroids, ribavirin, interferon or other RNA inhibitors [14].

5. Conclusions

The 2018 West Nile outbreak in FVG has been characterized by extraordinary high rates of WNV infections. The atypicality of the presentation of some of the WNNV forms highlights the necessity to stay vigilant and suspect the diagnosis when confronted with neurological symptoms. Given that no treatment has shown real efficacy, there is no consensus about the appropriate treatment. With regards to all the mosquito-borne infections, prevention measures, including personal protection and mosquito control programs, are necessary.

Author Contributions: Design of the study: M.B., M.P., N.C.; Data retrieval: all authors; Data analysis: M.B., M.P.; Writing of the manuscript: M.P., N.C.; Approval of final manuscript: all authors. All authors have read and agreed to the published version of the manuscript.

Funding: This research received no external funding.

Conflicts of Interest: The authors declare no conflict of interest.

References

1. Prevention CfDCa: West Nile Fever Transmission Season, Weekly Updates. Available online: http://ecdc.europa.eu/en/west-nile-fever/surveillance-and-disease-data/disease-data-ecdc (accessed on 20 January 2020).
2. Haussig, J.M.; Young, J.J.; Gossner, C.M.; Mezei, E.; Bella, A.; Sirbu, A.; Pervanidou, D.; Drakulovic, M.B.; Sudre, B. Early start of the West Nile fever transmission season 2018 in Europe. *Euro Surveill. Bull. Eur. Sur Les Mal. Transm. Eur. Commun. Dis. Bull.* **2018**, *23*. [CrossRef] [PubMed]
3. Epicentro. *Sorveglianza Integrata Del Wn E Usutuvirus*; Epicentro: Albany, NY, USA, 2018.
4. Sanchez, M.D.; Pierson, T.C.; McAllister, D.; Hanna, S.L.; Puffer, B.A.; Valentine, L.E.; Murtadha, M.M.; Hoxie, J.A.; Doms, R.W. Characterization of neutralizing antibodies to West Nile virus. *Virology* **2005**, *336*, 70–82. [CrossRef] [PubMed]
5. Levi, M.E. West Nile virus infection in the immunocompromised patient. *Curr. Infect. Dis. Rep.* **2013**, *15*, 478–485. [CrossRef] [PubMed]
6. Sambri, V.; Capobianchi, M.R.; Cavrini, F.; Charrel, R.; Donoso-Mantke, O.; Escadafal, C.; Franco, L.; Gaibani, P.; Gould, E.A.; Niedrig, M.; et al. Diagnosis of West Nile virus human infections: Overview and proposal of diagnostic protocols considering the results of external quality assessment studies. *Viruses* **2013**, *5*, 2329–2348. [CrossRef]
7. Lustig, Y.; Mannasse, B.; Koren, R.; Katz-Likvornik, S.; Hindiyeh, M.; Mandelboim, M.; Dovrat, S.; Sofer, D.; Mendelson, E. Superiority of West Nile virus RNA detection in whole blood for diagnosis of acute infection. *J. Clin. Microbiol.* **2016**, *54*, 2294–2297. [CrossRef] [PubMed]
8. Barzon, L.; Pacenti, M.; Franchin, E.; Pagni, S.; Martello, T.; Cattai, M.; Cusinato, R.; Palu, G. Excretion of West Nile virus in urine during acute infection. *J. Infect. Dis.* **2013**, *208*, 1086–1092. [CrossRef] [PubMed]
9. Centers for Disease Control and Prevention DoV-BD. *West Nile Virus in the United States: Guidelines for Surveillance, Prevention, and Control*; Centers for Disease Control and Prevention DoV-BD: Atlanta, GA, USA, 2018.
10. Granwehr, B.P.; Lillibridge, K.M.; Higgs, S.; Mason, P.W.; Aronson, J.F.; Campbell, G.A.; Barrett, A.D. West Nile virus: Where are we now? *Lancet Infect. Dis.* **2004**, *4*, 547–556. [CrossRef]
11. Petersen, L.R.; Brault, A.C.; Nasci, R.S. West Nile virus: Review of the literature. *JAMA* **2013**, *310*, 308–315. [CrossRef] [PubMed]
12. Alexander, J.J.; Lasky, A.S.; Graf, W.D. Stroke associated with central nervous system vasculitis after West Nile virus infection. *J. Child Neurol.* **2006**, *21*, 623–625. [CrossRef] [PubMed]
13. Radu, R.A.; Terecoasa, E.O.; Ene, A.; Bajenaru, O.A.; Tiu, C. Opsoclonus-myoclonus syndrome associated with West-Nile virus infection: Case report and review of the literature. *Front. Neurol.* **2018**, *9*, 864. [CrossRef] [PubMed]
14. D'Aes, T.; Marien, P. Cognitive and affective disturbances following focal brainstem lesions: A review and report of three cases. *Cerebellum* **2015**, *14*, 317–340. [CrossRef] [PubMed]
15. Harroud, A.; Almutlaq, A.; Pellerin, D.; Paz, D.; Linnell, G.J.; Gendron, D. West Nile virus-associated vasculitis and intracranial hemorrhage. *Neurol. Neuroimmunol. Neuroinflamm.* **2020**, *7*. [CrossRef] [PubMed]
16. Lowe, L.H.; Morello, F.P.; Jackson, M.A.; Lasky, A. Application of transcranial Doppler sonography in children with acute neurologic events due to primary cerebral and West Nile vasculitis. *AJNR Am. J. Neuroradiol.* **2005**, *26*, 1698–1701. [PubMed]
17. Maximova, O.A.; Pletnev, A.G. Flaviviruses and the central nervous system: Revisiting neuropathological concepts. *Annu. Rev. Virol.* **2018**, *5*, 255–272. [CrossRef] [PubMed]

18. Morrey, J.D.; Olsen, A.L.; Siddharthan, V.; Motter, N.E.; Wang, H.; Taro, B.S.; Chen, D.; Ruffner, D.; Hall, J.O. Increased blood-brain barrier permeability is not a primary determinant for lethality of West Nile virus infection in rodents. *J. Gen. Virol.* **2008**, *89*, 467–473. [CrossRef] [PubMed]
19. Suthar, M.S.; Diamond, M.S.; Gale, M., Jr. West Nile virus infection and immunity. *Nat. Rev. Microbiol.* **2013**, *11*, 115–128. [CrossRef] [PubMed]

Tropical Medicine and Infectious Disease

Letter

Letter to the Editor: Venezuelan Equine Encephalitis virus 1B Invasion and Epidemic Control—South Texas, 1971

Robert G. McLean

Retired, Research Program Manager, Wildlife Diseases Program, National Wildlife Research Center, Wildlife Services, USDA, APHIS, Fort Collins, CO 80521, USA; robertmclean285@gmail.com; Tel.: +01-970-391-1546

Received: 28 May 2020; Accepted: 17 June 2020; Published: 22 June 2020

Abstract: The epidemic strain of Venezuelan equine encephalitis virus (VEE) 1B invaded south Texas in 1971. The success of the eventual containment and control of the virus invasion was the early recognition and immediate detection, cooperation, coordination, and participation among multiple federal agencies. There were 4739 wild vertebrate animals trapped on a ranch in the area with only 1 VEE virus isolation from a Virgina opossum (Didelphis virginiana). A large number of mosquitoes were also collected on the ranch and tested, resulting in 240 VEE virus isolations. Virus isolations were obtained from 58% of the 33 equines tested. Wild vertebrates did not play a significant role in the outbreak.

Keywords: Venezuelan equine encephalitis virus; outbreak; arbovirus; interagency; response; coordination; mosquito; vaccine; emerging virus

---

1. Editor's Preface

In the letter below, Dr. McLean provides a first-hand account of the invasion of Venezuelan equine encephalitis virus 1B into South Texas. This invasion was met with a massive, coordinated, inter-agency response, which ultimately ended the outbreak. We believe this event represents the sole example to date in which invasion of an exotic zoonotic arbovirus into the United States has been eradicated, and exemplifies the interagency cooperation necessary to achieve such a successful outcome. During a time where we can expect the continued emergence of viruses that threaten human and animal health, such lessons from the field can be both informative and encouraging. Dr. McLean has led a very distinguished career, serving in leadership positions in both the Center for Disease Control (CDC) and the United States Department Agriculture (USDA), and we value his insight and perspective on this historic outbreak.

Dear Editor

The epidemic strain of Venezuelan equine encephalitis (VEE) 1B was responsible for epidemics in humans and equines in Ecuador in January 1969, and was subsequently introduced into coastal Guatemala and El Salvador (possibly in infected equines) in the summer of 1969. From there, the virus spread south to Honduras and Nicaragua, and north to Southern Mexico by November 1969. In 1970, the virus spread throughout those countries and north along the Gulf coast of Mexico toward Tampico, where it resulted in an epidemic there in March–April 1971. Descriptions of this outbreak have been published [1–4].

This continued northward progression alerted USDA Animal Plant Health Inspection Service (APHIS) and CDC to the likelihood of an invasion into the U.S. This impending hazard to the U.S. initiated the two agencies to start working together to develop contingency plans to address the hazard. A number of meetings were held among their staff and invited experts from other federal and state agencies and universities to develop strategies for prevention and control as well as to determine what investigations were needed. The U.S. Medical Research Institute of Infectious Diseases already had a VEE vaccine in their stock but it was not FDA/APHIS approved. However, it was made available to the team of investigators if they signed a voluntary release to protect them in Texas from VEE-1B virus infections that are quite debilitating in humans and kills equines. We were all willing to receive this unapproved (by FDA) vaccine so we could perform our duties in Texas. APHIS mobilized a number of teams of veterinarians, technicians, and administrators; CDC mobilized physicians, epidemiologists, entomologists (Dr. Vern Newhouse as leader), virologists (Dr. Charles Calisher as leader), wildlife biologists (Dr. Robert McLean as leader), and technicians for each. Counterpart staff from public health and other state agencies in Texas joined in the effort.

APHIS and CDC administrators and senior staff met in south Texas to develop local plans and establish administrative facilities and local and state connections and coordination. They were waiting for the first confirmed identification of VEE-1B virus in Texas (USA), whether equine, humans, or possibly mosquitoes or vertebrate animals. They could not enter Mexico to sample or test animals or humans to confirm the virus presence in nearby Northeast Mexico (across the Rio Grande river), and had to wait for it to spread into Texas. Arrangements were made to secure enough hotel and working space for the team members. All of the space at a motel in south Texas that was convenient for the potential affected area was reserved, and all of us later stayed there. We also used rooms to contain refrigerators and freezers, store equipment and supplies, setup a laboratory, and have meetings. It was the local epidemic investigation headquarters. APHIS had a large number of veterinarians there to deal with the investigation and control of VEE-1B virus in the equine population.

The team leaders selected staff to help conduct investigations, collected equipment and supplies and arranged for shipment to Southeast Texas, reserved vehicles and hotel reservations, and waited for the orders to proceed after the first VEE-1B virus detection in south Texas near the U.S.-Mexico border was confirmed. The first detection was a sick equine near the border. All of the teams were quickly mobilized to South Texas.

APHIS brought multiple teams of veterinarians for field investigations of affected equines and to establish quarantine and equine movement restrictions. CDC quickly setup a laboratory in the hotel to process and test blood samples for rapid detection of virus and antibody for infected equines and later humans after CDC epidemiologists began investigating human cases. All field supplies for all investigations of equine, human, mosquitoes, and wildlife were stored and distributed from that hotel, which was the headquarters for all of the entire U.S. government effort in South Texas. Coordination with state and local officials involved in the outbreak investigations was conducted, as well as with the press, from there.

Mosquito control efforts were developed, organized, and conducted locally, then broadened as the epidemic zone expanded. Finally, the U.S. Air Force fleet of large insecticide spray aircraft were deployed as the epidemic spread widely threatening larger areas of Texas to the north [5].

Because equines were the primary host for amplification, a big achievement for the control of the epidemic and the protection of equines was the very rapid emergency approval of the VEE vaccine by USDA and the production of the military VEE vaccine for use in equines by a private vaccine company. The vaccine was made available for use by private veterinarians throughout the state of Texas. One industrious veterinarian acquired state approval and distributed the vaccine by a rented private plane to veterinarians throughout the state. Control of the epidemic and advancement northward was achieved, but there were about 1800 equine cases in the southern half of the state by the time the epidemic was stopped more than two months later.

2. Wildlife Investigations

At this time, I was the only wildlife biologist/epidemiologist at CDC and was in the Rabies branch of the epidemiology program. I was requested and agreed to head the wildlife investigations and selected some team members to assist. Luckily, I had just hired a biologist from south Texas who had worked for CDC at the public health unit at the border there that was closing and who grew up in that area. He had many contacts there and helped arrange in selecting a sampling location near the border on a large cattle ranch with great habitats for wildlife. The rancher allowed us free access to all of the ranch and provided help and guidance from his ranch manager.

We set a large number of traps for rodents; smaller numbers of traps for larger mammals such as raccoons, opossums, and rabbits; and leg-hold traps for carnivores and captured tortoises by hand throughout the ranch. At the same time, we set up mist nets for birds throughout the area and operated some mist nets at night for bats. Canon nets were set up on the beach on South Padre Island to capture sea gulls. Mosquito light traps were placed within the lines of mist nets and mammal traps to obtain a comprehensive sampling of vector and vertebrate populations for a more thorough effort to detect the presence of VEE 1B virus. Entomologist Dr. Newhouse from CDC was the leader of the mosquito sampling effort, although we all worked together to conduct prolonged and intense field investigations during a multi-week period in August—September 1971.

Follow-up investigations were conducted multiple times at these sampling sites by the vertebrate and vector teams during the next year to ensure that VEE 1B virus did not become established in South Texas.

3. Laboratory Testing

Some of the testing of blood specimens from vertebrate animals, humans and processed mosquitoes were conducted locally to obtain rapid results for the epidemiologists and mosquito control teams who were trying to stop the epidemic expansion.

The majority of the testing was conducted at CDC in Atlanta. Virus isolation was performed by inoculating 0.02 mL of each sample by the intracerebral route into Swiss albino suckling mice. Virus isolates were identified by microtiter complement fixation tests. Virus neutralizing antibody in the vertebrate and human sera were tested in duck embryo cell culture by the plaque reduction neutralization test using VEE-TC-83 vaccine virus.

4. Conclusion

This was a massive, coordinated effort to prepare for the impending invasion of the epidemic 1B strain of VEE virus into the USA at the Texas border. The success of the eventual containment and control of the virus invasion was the early recognition and immediate detection, cooperation, coordination, and participation among multiple federal agencies of USDA, CDC, U.S. Military, National Park Service, Department of Interior), multiple Texas state departments, and local agencies and their staff with valuable input from individual scientific experts. The leaders of these agencies were on the ground in South Texas to initiate this massive, coordinated federal/state response. A well-planned response was developed, and multiple teams were formed to investigate specific health aspects in the human and equine populations, and the mosquito vector and wild vertebrate host involvement. The effective control of virus transmission was initiated quickly through broad equine vaccination and quarantine and intensive mosquito control that prevented the epidemic spread to Northern Texas and surrounding states. The close cooperation and joint leadership among the federal, state, and local officials and the dedicated and exhaustive efforts by the scientists and public health officials and their staff was key for the success of this dramatic disease control effort. It was a good example for dealing with foreign disease invasion and control. A big advantage for the control of this invasion and impending epidemic in Texas was the prior knowledge about the virus strain, history of its origin, and movement from South America to Central America and northward along the Gulf coast of Mexico, and the information on the potential domestic and wild vertebrate hosts and vectors that would be affected.

Funding: Preparation of this work was supported in part by the Department of Homeland Security (DHS) Science and Technology Directorate (S&T) under agreement # HSHQDC-17-C-B0003. The viewpoints described herein represent those of the author and not necessarily their institutions or funding agency.

Acknowledgments: This manuscript was formatted by R.C.K.

Conflicts of Interest: The author declares no conflict of interest. The funders had no role in the design of the study; in the collection, analyses, or interpretation of data; in the writing of the manuscript; or in the decision to publish the results.

References

1. Sudia, W.D.; Newhouse, V.F. Venezuelan equine encephalitis in texas information reports 1971. *Mosq. News* **1971**, *31*, 350–351.
2. Zehmer, R.B.; Dean, P.B.; Sudia, W.D.; Calisher, C.H.; Sather, G.E.; Parker, R.L. Venezuelan, Equine Encephalitis Epidemic in Texas, 1971. *Health Serv. Rep.* **1974**, *89*, 278–282. [CrossRef] [PubMed]
3. Sudia, W.D.; McLean, R.G.; Newhouse, V.F.; Johnston, J.G., Jr.; Miller, D.L.; Trevino, H.A.; Bowen, G.S.; Sather, G.E. Epidemic Venezuelan equine encephalitis in North America in 1971: Vertebrate field studies. *Am. J. Epidemiol.* **1975**, *101*, 36–50. [CrossRef] [PubMed]
4. Sudia, W.D.; Newhouse, V.F.; Beadle, I.D.; Miller, D.L.; Johnston, J.G., Jr.; Young, R.; Calisher, C.H.; Maness, K. Epidemic Venezuelan equineencephalitis in North America in 1971: Vector studies. *Am. J. Epidemiol.* **1975**, *101*, 17–35. [CrossRef] [PubMed]
5. Pinkovsky, D.D. United States Air Force aerial spray activities in operation combat VEE. *Mosq. News* **1972**, *328*, 332–337.

Tropical Medicine and Infectious Disease

Perspective

Emergence of Arboviruses in the United States: The Boom and Bust of Funding, Innovation, and Capacity

Rebekah C. Kading [1,*], Lee W. Cohnstaedt [2], Ken Fall [3] and Gabriel L. Hamer [4]

1 Department of Microbiology, Colorado State University, Immunology and Pathology, Fort Collins, CO 80523, USA
2 United States Department of Agriculture, Arthropod-borne Animal Diseases Research Unit, Agricultural Research Service, Manhattan, KS 66502, USA; lee.cohnstaedt@usda.gov
3 BioQuip Products, Rancho Dominguez, CA 90220, USA; ken@bioquip.com
4 Department of Entomology, College of Agriculture and Life Sciences, Texas A&M University, College Station, TX 77843, USA; ghamer@tamu.edu
* Correspondence: rebekah.kading@colostate.edu

Received: 29 April 2020; Accepted: 4 June 2020; Published: 6 June 2020

Abstract: Mosquito-borne viruses will continue to emerge and generate a significant public health burden around the globe. Here, we provide a longitudinal perspective on how the emergence of mosquito-borne viruses in the Americas has triggered reactionary funding by sponsored agencies, stimulating a number of publications, innovative development of traps, and augmented capacity. We discuss the return on investment (ROI) from the oscillation in federal funding that influences demand for surveillance and control traps and leads to innovation and research productivity.

Keywords: mosquito; emerging virus; outbreak; surveillance; trap; *Aedes*; *Culex*; Zika virus; West Nile virus

---

1. Introduction

While outbreaks of infectious diseases are devastating to human and animal populations, surges in private and public funding for outbreak pathogens have positively influenced research and innovation. Contemporary outbreak response efforts now incorporate digital media; electronic surveillance tools; mathematical modelling; sequencing; and teams of experts that include anthropologists, social scientists, and other diverse disciplinarians [1]. Outbreak stimulus funding leads to increased scientific productivity on outbreak pathogens, and incites innovation [2]. For example, an innovation workshop convened after an anthrax bioterrorism attack in 2001, to discuss new surveillance and pathogen detection approaches [3]. The Ebola Grand Challenge program, funded by the US Centers for Disease Control and Prevention (CDC), United States Agency for International Development (USAID), the United States Executive Office, and the Department of Defense, provided financial backing to 14 innovative projects to improve the response to Ebola outbreaks. These projects included protective suits, health care and technology solutions, and decontaminants [4]. The USAID Zika Grand Challenge program in 2016 received over 900 applications and provided 30 million dollars to innovative ideas to address Zika virus transmission [5]. Currently, during the COVID-19 pandemic, innovations are emerging such as new ventilators, drones delivering medical supplies, and the use of artificial intelligence in medicine [6–9]. Ideally, these scientific advances will aid in combatting the current epidemic as well as fuel innovation in preparedness for future emerging disease events, leading to return on investment (ROI).

The introduction and spread of numerous mosquito-borne viruses around the globe has also shepherded a wave of innovation to improve upon industry-standard sampling techniques that more efficiently capture the target vector species, physiological cohort, or virus-infected population of vectors [10]. Ramirez et al. [10] recently reviewed traditional surveillance approaches in the context of novel innovations that have advanced surveillance capacity, including pathogen surveillance from sugar feeding vectors, next-generation sequencing (eDNA (environmental DNA) for presence or absence of vectors in habitats), and xenosurveillance for vectors and pathogens. Additionally, modernized sampling strategies have also been expanded to incorporate technical advancements in such areas as infrared scanning, citizen science, and drones [11–14]. Driving the introduction and incorporation of novel trapping approaches to surveillance programs is the persistent emergence of mosquito-borne pathogens.

2. Arbovirus Emergence in the U.S.

Many arboviruses cause a significant disease burden to the public each year. West Nile virus (WNV) (introduced in 1999), St. Louis encephalitis virus (SLEV), Jamestown Canyon virus (JCV), La Crosse encephalitis virus (LACV), Eastern equine encephalitis virus (EEEV), and Powassan (POWV, tick-borne) are tracked through the CDC database, ArboNET [15]. In the last several years, incidence of JCV and POWV has been increasing [15–18]. Additionally, a multistate outbreak of EEEV occurred in the northeastern US in 2019, in which the 34 reported cases far exceeded the historical average of around eight cases/year [19]. Through the first half of the 20th century, Western equine encephalitis virus (WEEV) was also widespread and caused significant morbidity and mortality in humans and horses; however, this virus has largely disappeared from most areas over the past several decades for unconfirmed reasons, with the last human case in the United States reported in 1999 [20,21].

In 2013, chikungunya (CHIKV) was first reported in the Americas, followed by Zika virus (ZIKV) two years later [22,23]. These virus invasions came during a time of increasing dengue virus (DENV) incidence, range expansion, and replacement of dominant serotypes in different geographic areas [24–26]. The Pan American Health Organization (PAHO) reported a 30% increase in the number of dengue cases, from 7,641,334 between 2001 and 2010 to 10,851,043 between 2011 and 2017 [26,27]. Intensive transmission of *Aedes*-borne viruses in the Americas coupled with the expanding range of *Ae. aegypti* mosquitoes presents a continual threat of these viruses gaining a foothold in the United States as well [28–30]. To date, cases of DENV, ZIKV, and CHIKV in the United States have predominately been traveler-associated, with the exception of local transmission of ZIKV in Florida and Texas during 2017 [31,32].

3. Federal Funding and Publications in Response to Arboviral Emergence

When a mosquito-borne virus rapidly spreads around the world and becomes a public health emergency of international concern, federal agencies allocate substantial funds to fight the virus. In the U.S., Congress approves these funding packages, such as the $1.1 billion for Zika virus in September, 2016 [33]. This stimulus funding provides agencies such as the National Institutes of Health (NIH), the Centers for Disease Control and Prevention (CDC), and the National Science Foundation the resources to allocate funds to public health agencies and researchers to enhance outbreak surveillance and response efforts. As a proxy for this federal spending allocated to different emerging arboviruses, we used the NIH RePORTER search query to estimate the funding of projects related to the emergence of WNV, CHIKV, and ZIKV. We used the key words of "West Nile virus", "chikungunya virus", and "Zika virus" in the search feature which queries all project titles, abstracts, and scientific terms. To compare these invasive mosquito-borne viruses to those endemic to the United States, we also included three endemic mosquito-borne viruses that have resulted in continual transmission, including human disease, in the last several decades. These additional virus searches were performed for "Eastern equine encephalitis virus", "St Louis encephalitis virus", and "La Crosse encephalitis virus". The search for St. Louis encephalitis virus returned more projects with "St" than with "Saint". The search was performed on 16 May 2020, and we included all years from 1985–2020 in each search. NIH-funded projects have a record for each fiscal year of a project (e.g., a 5-year NIH R01 has five records). The projects include international and national investigators and project locations. The search for WNV resulted in 2803 records for a total of $1,493,811,562 in funding; CHIKV matched 960 records for $477,117,414 in funding; and ZIKV matched 1332 records for $1,164,293,217 in funding (Figure 1). The search for EEEV matched 310 records for a total of $165,906,376 in funding; SLEV matched 143 records for $59,372,038 in funding; and LACV matched 126 records for $41,122,419 in funding (Figure 2). The rise in NIH-funded projects for the invasive arboviruses rose quickly following their introduction into the U.S., especially for ZIKV. The NIH-funding for endemic arboviruses was steadier over the 35 time year period, especially for LACV.

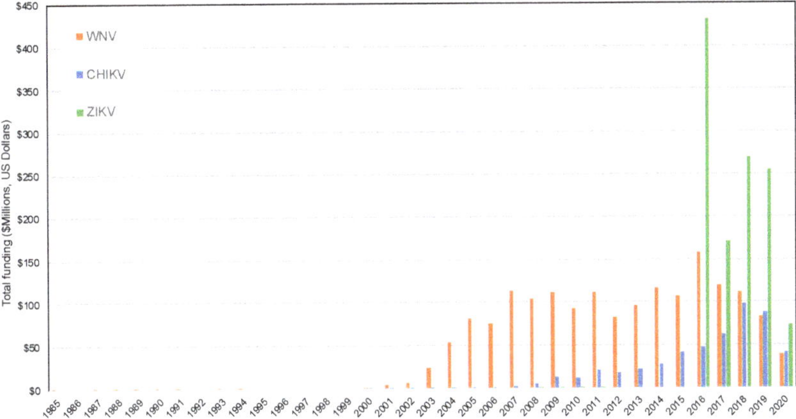

Figure 1. Total funding by the National Institutes of Health based on the National Institutes of Health (NIH) Reporter search query of projects matching West Nile virus (WNV), chikungunya virus (CHIKV), and Zika virus (ZIKV), from 1985 to 2020. Expenditures in 2020 are incomplete as search was conducted on 16 May 2020. Graphed values were corrected for inflation by the Consumer Price Index referenced to 2019 provided by the U.S. Bureau of Labor Statistics.

In response to WNV and ZIKV, the CDC supports detection and response capacity to local agencies by providing states Epidemiology and Laboratory Capacity (ELC) cooperative agreements with all 50 states and US territories. For the first five years following the introduction of WNV into the U.S., the CDC issued ELC funds which were reduced in 2004. A survey by the Council of State and Territorial Epidemiologists in 2005 showed several enhancements to surveillance, laboratory, and control capacity that were made possible as the result of ELC funding [34]. However, after yearly reductions in ELC funds, a survey in 2013 revealed many of these programs had reduced or lost surveillance and response measures [35]. Despite this lack of sustained funding and reduced infrastructure, WNV continues to have periodic large epidemics [36]. The advent of CHIKV and ZIKV to the Americas caught many U.S local, state, and federal agencies off guard with a lack of knowledge of, lack of surveillance of, or inability to control *Aedes* (*Stegomyia*) mosquitoes. Once again, along with the NIH who spent over $400 million in 2016 on ZIKV-related projects, the CDC administered $97 million in supplementary Zika ELC stimulus funds to states to help programs re-bound and shift from a *Culex*-centric focus to *Aedes* [37]. This pulse in funding helped improve the surveillance and control of *Aedes* mosquitoes, although the fading of Zika virus in the national news and the reduction of these ELC funds [38] has resulted in many programs again reducing their capacity.

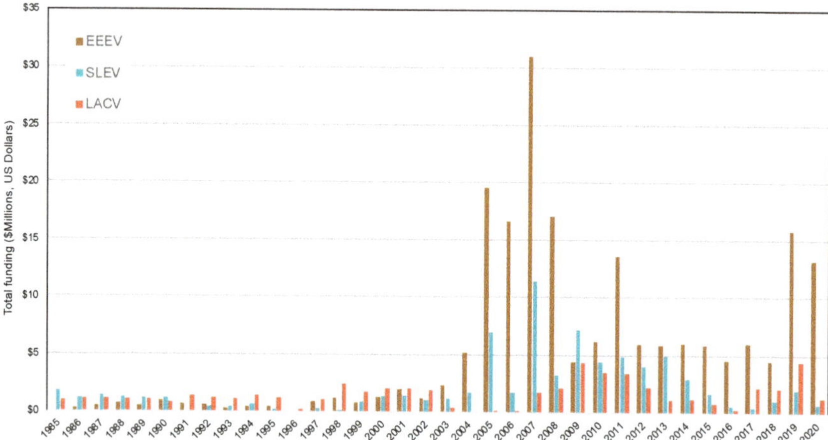

Figure 2. Total funding by the National Institutes of Health based on the NIH Reporter search query of projects matching Eastern equine encephalitis virus (EEEV), St Louis encephalitis virus (SLEV), and La Crosse encephalitis virus (LACV) from 1985 to 2020. Expenditures in 2020 are incomplete as search was conducted on 16 May 2020. Graphed values were corrected for inflation by the Consumer Price Index referenced to 2019 provided by the US Bureau of Labor Statistics.

Along with the pulse in funding and attention by media and the public, surges in peer-reviewed publications occur following the emergences of arboviruses. To quantify these publications, we performed a "Basic Search" of the Web of Science Core Collection for the key words of "West Nile virus", "chikungunya virus", and "Zika virus". As with the NIH RePORTER search, we included "Eastern equine encephalitis virus", "St Louis encephalitis virus", and "La Crosse encephalitis virus". The search for St. Louis encephalitis virus returned more publications with "St" than with "Saint". The year range for the search was 1985 to 2020, and the search was conducted on 16 May 2020. The results yielded 11,868 records for WNV; 5088 for CHIKV; 8074 for ZIKV; 634 for EEEV; 604 for SLEV; and 238 for LACV (Figures 3 and 4). The yearly pattern in publications for these three arboviruses closely matches the total funding by NIH, although publications reported on Web of Science include publications by international authors funded by different international sponsors. Publications including the endemic arboviruses as keyword searches have slowly increased over the last 35 years, but the scale of the increase in publications for invasive arboviruses is much greater.

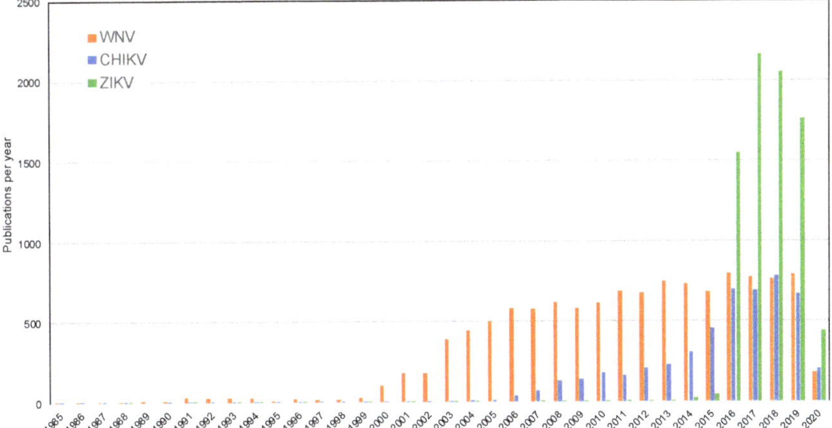

Figure 3. Results of a Basic Search of the Web of Science Core Collection for West Nile virus (WNV), chikungunya virus (CHIKV), and Zika virus (ZIKV), from 1985 to 2020. Publications in 2020 are incomplete as search was conducted on 16 May 2020.

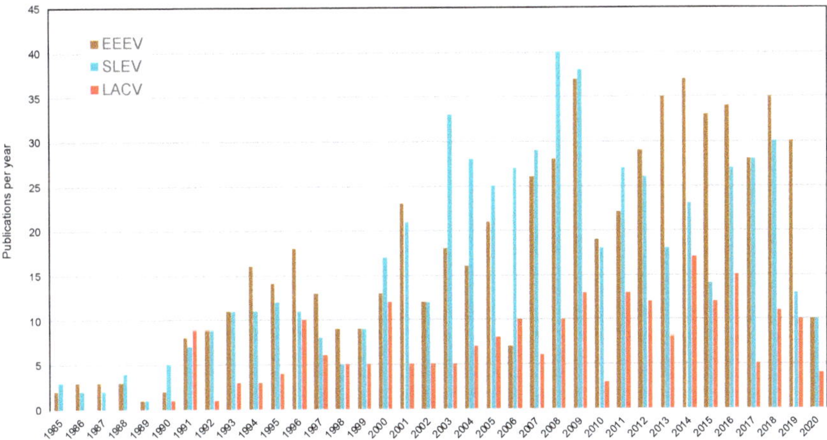

Figure 4. Results of a Basic Search of the Web of Science Core Collection for eastern equine encephalitis virus (EEEV), St. Louis encephalitis virus (SLEV), and La Crosse encephalitis virus (LACV) from 1985 to 2020. Publications in 2020 are incomplete as search was conducted on 16 May 2020.

4. Innovation of Mosquito Surveillance Tools

The evolution of mosquito trapping tools targeting *Culex* and *Aedes* mosquitoes has followed suit, being driven by the emergence of these viruses, the resulting resources generated by federal agencies, and consumer demand. To visualize trends in trap usage over time, we evaluated BioQuip Products sales data in each main trap category between 2003 and 2019 (Figure 5). As the United States has experienced emergences and threats of first *Culex* and now also *Aedes*-borne viruses, innovations to traditional mosquito trapping tools targeting these vector groups have arisen, with the proportion of sales in different categories fluctuating over time in response to arbovirus outbreak and funding availability (Figure 1). The diversity of traps sold by BioQuip Products also increased substantially from two categories in 2003 to nine categories by 2014 (Figure 3) as new innovations came to market. Undoubtedly, the surge of *Aedes*-borne viruses in particular during this time period has influenced the development and sales of novel traps targeting *Ae. aegypti* mosquitoes.

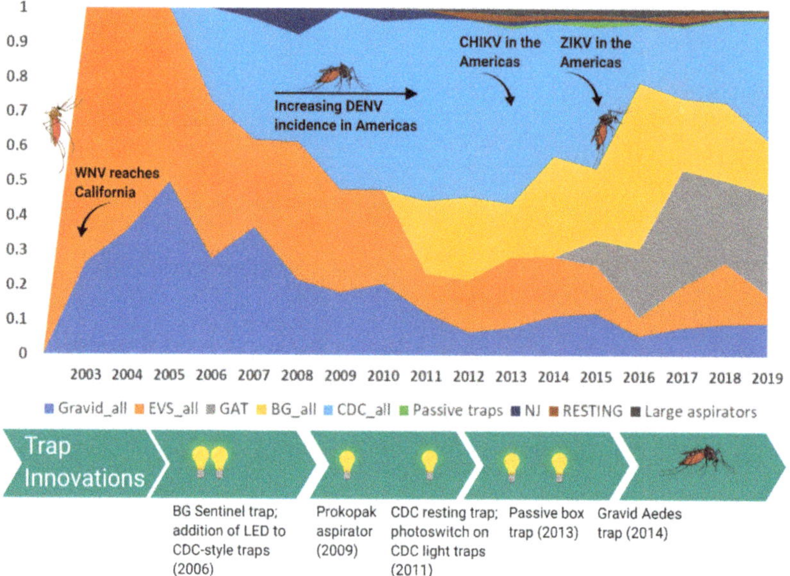

Figure 5. Proportion of traps sold per year by BioQuip Products. Product numbers represented included: Gravid traps (2800, 2800S), BG traps (2880, 2883), NJ light trap (2856), EVS traps (2801A, 2780, 2780NS1, 2780NS2), Centers for Disease Control and Prevention (CDC) traps (2848, 2770, 2836BQ, 2836BQX), resting traps (2799), Gravid Aedes Traps (2797), passive traps (2887, 2887P), large aspirators (2888A, 2846). Inset: Proportion of trap types sold per year by BioQuip Products. Product numbers represented included: Gravid traps (2800, 2800S), BG traps (2880, 2883), NJ light trap (2856, 2857, 2858), EVS traps (2801A, 2780, 2780NS1, 2780NS2), CDC traps (2848, 2770, 2836BQ, 2836BQX), resting traps (2799), Gravid Aedes Traps (2797), passive traps (2887, 2887P), and large aspirators (2888A, 2846).

4.1. *Culex-Borne Virus Surveillance*

The CDC light trap has become an industry standard for collection of mosquitoes for the purposes of arbovirus surveillance. The similar EVS (Encephalitis Vector Survey) trap [39] is also routinely used for arbovirus surveillance and operates on a similar principle as the CDC light trap using light and carbon dioxide (CO_2) as attractants. Between 2003 and 2010, trap sales were dominated by suction traps using light and CO_2 as attractants, and gravid traps, largely for *Culex* vectors of WNV and SLEV. Due to the sustained threat of WNV year after year [36], traps targeting *Culex* vectors are expected to remain a mainstay in the industry.

Several specific innovations have either improved the functionality of the standard light trap or targeted different physiological cohorts of mosquitoes. In 2006, BioQuip released the first commercially available light-emitting-diode (LED) CDC-style light trap with customizable colors (Figure 3). Bayonet mounted LED chips were also available to replace the white incandescent lights in existing CDC style light traps. The passive box trap offered an alternative to CDC light traps, by providing a CO_2-baited trap that was not reliant on a power source and could be used to sample arbovirus vectors in more remote areas [40]. A collapsible passive trap was also developed to address portability issues [41]. A photo switch option also added to light traps, allowing researchers the flexibility to set traps when convenient and conserve battery power. Finally, the CDC resting trap was introduced in 2011 for the purpose of collecting blood-engorged mosquitoes to determine vertebrate host utilization [42]. Passive and resting traps represent a smaller proportion of sales over the last several years, suggesting their utility may be serving the research community as opposed to being integrated into large-scale operational surveillance activities.

4.2. Aedes-Borne Virus Surveillance

Surveillance for viruses transmitted by *Ae. aegypti* mosquitoes requires very different tools and strategies than for those used for *Culex*-borne viruses. These urbanized, day-biting, container-breeding mosquitoes are generally not attracted to the suction and gravid traps that are so efficient at catching *Culex* mosquitoes [43]. The growing need for traps that effectively collected *Aedes* mosquitoes led to the innovation of the BG Sentinel Mosquito Trap in 2006 (Biogents AG, Regensburg, Germany) [44], among others. BG sentinel traps began selling at BioQuip in 2011, and sales comprised approximately 50% of the total traps sold by BioQuip in 2016 following the introduction of CHIKV and ZIKV to the Americas (Figure 3). Similarly, gravid *Aedes* traps (GAT) [45,46] were introduced between 2013 and 2014. Conceptually based on traditional gravid traps [47], these traps do not require electricity to function. GAT sales quickly grew to approximately 30% of the traps sold by BioQuip in 2016 (Figure 5). The InsectaZooka© and the Prokopak aspirator [48] are lightweight, portable field aspirators developed as an alternative to backpack models. These devices facilitate the collection of resting and engorged mosquitoes from indoor and outdoor resting sites.

5. Conclusions

The emergence of mosquito-borne viruses has occurred repeatedly in recent decades, and studies predict that these emergences will continue. Much of our response to these events is reactionary, triggered by the increase in attention, funding, publication, innovation, and preventive measures for public health. The long-term impact or return on investment of outbreak spending is evidenced by scientific advancements (publications) and innovation, but we advocate for a more sustainable, economical, and effective approach by minimizing the oscillations of boom and bust in funding and capacity for mosquito-borne viruses. Our goal should be to optimize the cost-effectiveness of budgetary spending by adopting resource allocation for biosecurity threats that maximize benefits while minimizing the total cost given anticipated expenditures incurred in the event of mosquito-borne viral outbreaks [49,50].

Author Contributions: Conceptualization, G.L.H., R.C.K., and L.W.C.; data curation, K.F.; writing—original draft preparation, R.C.K.; writing—review and editing, R.C.K., G.L.H., L.W.C., and K.F.; visualization, R.C.K., G.L.H., and L.W.C.; project administration, G.L.H.; funding acquisition, G.L.H. All authors have read and agreed to the published version of the manuscript.

Funding: This research was funded in part by the Department of Homeland Security (DHS) Science and Technology Directorate (S&T) under agreement # HSHQDC-17-C-B0003.

Acknowledgments: The viewpoints described herein represent those of the authors and not necessarily their institutions or funding agency.

Conflicts of Interest: The authors declare no conflict of interest. The funders had no role in the design of the study; in the collection, analyses, or interpretation of data; in the writing of the manuscript; or in the decision to publish the results.

References

1. Houlihan, C.F.; Whitworth, J.A. Outbreak science: Recent progress in the detection and response to outbreaks of infectious diseases. *Clin. Med.* **2019**, *19*, 140–144. [CrossRef] [PubMed]
2. Jayabalisingham, B.; Hessen, M.; James, C. Infectious Disease Outbreak Research: Insights and Trends 2020 [Webinar] In Scopus Webinar Series. Available online: https://www.brighttalk.com/webcast/13703/391874. (accessed on 1 April 2020).
3. Pavlin, J.A.; Mostashari, F.; Kortepeter, M.G.; Hynes, N.A.; Chotani, R.A.; Mikol, Y.B.; Ryan, M.A.K.; Neville, J.S.; Gantz, D.T.; Writer, J.V.; et al. Innovative Surveillance Methods for Rapid Detection of Disease Outbreaks and Bioterrorism: Results of an Interagency Workshop on Health Indicator Surveillance. *Am. J. Public Health* **2003**, *93*, 1230–1235. [CrossRef] [PubMed]
4. USAID. Fighting Ebola: A Grand Challenge for Development. Available online: http://www.ebolagrandchallenge.net (accessed on 7 April 2020).

5. USAID. Combating Zika | Grand Challenge for Development | U.S. Agency for International Development. Available online: https://www.usaid.gov/grandchallenges/zika (accessed on 25 April 2020).
6. Bashir, N. James Dyson Designed a New Ventilator in 10 days. He's Making 15,000 for the Pandemic Fight. Available online: https://www.cnn.com/2020/03/26/tech/dyson-ventilators-coronavirus/index.html (accessed on 7 April 2020).
7. Bhatt, K.; Pourmand, A.; Sikka, N. Targeted Applications of Unmanned Aerial Vehicles (Drones) in Telemedicine. *Telemed. E-Health* **2018**, *24*, 833–838. [CrossRef] [PubMed]
8. Roblin, S. Will Blood-Bearing Delivery Drones Transform Disaster Relief and Battlefield Medicine? Available online: https://www.forbes.com/sites/sebastienroblin/2019/10/22/will-blood-bearing-delivery-drones--transform-disaster-relief-and-battlefield-medicine/ (accessed on 7 April 2020).
9. Shapiro, G. *How Innovation Is Helping Mitigate the Coronavirus Threat*; STAT: Boston, MA, USA, 2020.
10. Ramírez, A.; Meyer, D.; Ritchie, S. Searching for the proverbial needle in a haystack: Advances in mosquito-borne arbovirus surveillance. *Parasit. Vectors* **2018**, *11*, 1–12. [CrossRef] [PubMed]
11. Cohnstaedt, L.; Ladner, J.; Campbell, L.; Busch, N.; Barrera, R. Determining Mosquito Distribution from Egg Data: The Role of the Citizen Scientist. *Am. Biol. Teach.* **2016**, *78*, 317–322. [CrossRef]
12. Esperança, P.M.; Blagborough, A.M.; Da, D.F.; Dowell, F.E.; Churcher, T.S. Detection of Plasmodium berghei infected Anopheles stephensi using near-infrared spectroscopy. *Parasit. Vectors* **2018**, *11*, 377. [CrossRef]
13. Fernandes, J.N.; Dos Santos, L.M.B.; Chouin-Carneiro, T.; Pavan, M.G.; Garcia, G.A.; David, M.R.; Beier, J.C.; Dowell, F.E.; Maciel-De-Freitas, R.; Sikulu-Lord, M.T. Rapid, noninvasive detection of Zika virus in mosquitoes by near-infrared spectroscopy. *Sci. Adv.* **2018**, *4*, eaat0496. [CrossRef]
14. Norris, D.; Ray, A.; Guda, T.; Keogh, E.; Singh, S.; Zhu, Y.; Debboun, M.; Reyna, M.; Vigilant, M.; Carpi, G.; et al. Project premonition project: Field trials of a robotic smart trap for mosquito identification and bionomics. *Am. J. Trop. Med. Hyg.* **2017**, *97*, 610–611.
15. Curren, E.J.; Lehman, J.; Kolsin, J.; Walker, W.L.; Martin, S.W.; Staples, J.E.; Hills, S.L.; Gould, C.V.; Rabe, I.B.; Fischer, M.; et al. West Nile Virus and Other Nationally Notifiable Arboviral Diseases—United States, 2017. *MMWR Morb. Mortal. Wkly. Rep.* **2018**, *67*, 1137–1142. [CrossRef]
16. Gaensbauer, J.T.; Lindsey, N.P.; Messacar, K.; Staples, J.E.; Fischer, M. Neuroinvasive arboviral disease in the United States: 2003 to 2012. *Pediatrics* **2014**, *134*, e642–e650. [CrossRef]
17. Krow-Lucal, E.R.; Lindsey, N.P.; Fischer, M.; Hills, S.L. Powassan Virus Disease in the United States, 2006–2016. *Vector Borne Zoonotic Dis.* **2018**, *18*, 286–290. [CrossRef] [PubMed]
18. Pastula, D.M.; Hoang Johnson, D.K.; White, J.L.; Dupuis, A.P.; Fischer, M.; Staples, J.E. Jamestown Canyon Virus Disease in the United States-2000–2013. *Am. J. Trop. Med. Hyg.* **2015**, *93*, 384–389. [CrossRef] [PubMed]
19. Lindsey, N.P.; Martin, S.W.; Staples, J.E.; Fischer, M. Notes from the Field: Multistate Outbreak of Eastern Equine Encephalitis Virus—United States, 2019. *MMWR Morb. Mortal. Wkly. Rep.* **2020**, *69*, 50–51. [CrossRef] [PubMed]
20. Adams, D.; Gallagher, K.; Jajosky, R.A.; Ward, J.; Sharp, P.; Anderson, W.; Abellera, J.; Aranas, A.; Mayes, M.; Wodajo, M.; et al. Summary of Notifiable Diseases—United States, 2010. *Morb. Mortal. Wkly. Rep.* **2010**, *59*, 1–111.
21. Robb, L.L.; Hartman, D.A.; Rice, L.; deMaria, J.; Bergren, N.A.; Borland, E.M.; Kading, R.C. Continued Evidence of Decline in the Enzootic Activity of Western Equine Encephalitis Virus in Colorado. *J. Med. Entomol.* **2019**, *56*, 584–588. [CrossRef]
22. Fischer, M.; Staples, J.E. Arboviral Diseases Branch, National Center for Emerging and Zoonotic Infectious Diseases, CDC Notes from the field: Chikungunya virus spreads in the Americas—Caribbean and South America, 2013–2014. *MMWR Morb. Mortal. Wkly. Rep.* **2014**, *63*, 500–501.
23. Zanluca, C.; de Melo, V.C.A.; Mosimann, A.L.P.; Santos, G.I.V.D.; Santos, C.N.D.D.; Luz, K. First report of autochthonous transmission of Zika virus in Brazil. *Mem. Inst. Oswaldo Cruz* **2015**, *110*, 569–572. [CrossRef]
24. Balmaseda, A.; Standish, K.; Mercado, J.C.; Matute, J.C.; Tellez, Y.; Saborío, S.; Hammond, S.N.; Nuñez, A.; Avilés, W.; Henn, M.R.; et al. Trends in patterns of dengue transmission over 4 years in a pediatric cohort study in Nicaragua. *J. Infect. Dis.* **2010**, *201*, 5–14. [CrossRef]
25. Brathwaite, D.O.; San Martín, J.L.; Montoya, R.H.; del Diego, J.; Zambrano, B.; Dayan, G.H. The history of dengue outbreaks in the Americas. *Am. J. Trop. Med. Hyg.* **2012**, *87*, 584–593. [CrossRef]

26. Salles, T.S.; da Encarnação Sá-Guimarães, T.; de Alvarenga, E.S.L.; Guimarães-Ribeiro, V.; de Meneses, M.D.F.; de Castro-Salles, P.F.; Dos Santos, C.R.; do Amaral Melo, A.C.; Soares, M.R.; Ferreira, D.F.; et al. History, epidemiology and diagnostics of dengue in the American and Brazilian contexts: A review. *Parasit. Vectors* **2018**, *11*, 264. [CrossRef]
27. PAHO. Number of Reported Cases of Dengue and Dengue Hemorrhagic Fever (DHF), Region of the Americas (by Country and Subregion) 1980–2018. Available online: https://www.paho.org/data/index.php/en/mnu-topics/indicadores-dengue-en/dengue-nacional-en/252-dengue-pais-ano-en.html (accessed on 7 April 2020).
28. Kraemer, M.U.G.; Reiner, R.C.; Brady, O.J.; Messina, J.P.; Gilbert, M.; Pigott, D.M.; Yi, D.; Johnson, K.; Earl, L.; Marczak, L.B.; et al. Past and future spread of the arbovirus vectors Aedes aegypti and Aedes albopictus. *Nat. Microbiol.* **2019**, *4*, 854–863. [CrossRef] [PubMed]
29. Pless, E.; Gloria-Soria, A.; Evans, B.R.; Kramer, V.; Bolling, B.G.; Tabachnick, W.J.; Powell, J.R. Multiple introductions of the dengue vector, Aedes aegypti, into California. *PLoS Negl. Trop. Dis.* **2017**, *11*, e0005718. [CrossRef] [PubMed]
30. Ryan, S.J.; Carlson, C.J.; Mordecai, E.A.; Johnson, L.R. Global expansion and redistribution of Aedes-borne virus transmission risk with climate change. *PLoS Negl. Trop. Dis.* **2019**, *13*, e0007213. [CrossRef] [PubMed]
31. Castro, L.A.; Fox, S.J.; Chen, X.; Liu, K.; Bellan, S.E.; Dimitrov, N.B.; Galvani, A.P.; Meyers, L.A. Assessing real-time Zika risk in the United States. *BMC Infect. Dis.* **2017**, *17*, 284. [CrossRef] [PubMed]
32. Florida State Health Department. Department of Health Daily Zika Update | Florida Department of Health. Available online: http://www.floridahealth.gov/newsroom/2017/01/012717-zika-update.html (accessed on 18 April 2020).
33. Kaplan, S. Congress Approves $1.1 Billion in Zika Funding. Available online: https://www.scientificamerican.com/article/congress-approves-1-1-billion-in-zika-funding/ (accessed on 15 April 2020).
34. MMWR. Assessing Capacity for Surveillance, Prevention, and Control of West Nile Virus Infection—United States, 1999 and 2004. Available online: https://www.cdc.gov/mmwr/preview/mmwrhtml/mm5506a2.htm (accessed on 15 April 2020).
35. Hadler, J.; Patel, D.; Bradley, K.; Hughes, J.; Blackmore, C.; Etkind, P.; Kan, L.; Getchell, J.; Blumenstock, J.; Engel, J. National Capacity for Surveillance, Prevention, and Control of West Nile Virus and Other Arbovirus Infections—United States, 2004 and 2012. *MMWR Morb. Mortal. Wkly. Rep.* **2014**, *63*, 281–284.
36. CDC. Statistics & Maps | West Nile Virus | CDC. Available online: https://www.cdc.gov/westnile/statsmaps/index.html (accessed on 18 April 2020).
37. CDC. CDC Awards Nearly $184 Million to Continue the Fight against Zika. Available online: https://www.cdc.gov/media/releases/2016/p1222-zika-funding.html (accessed on 15 April 2020).
38. CDC. ELC 5-year Focus, Funding, and Impact | DPEI | CDC. Available online: https://www.cdc.gov/ncezid/dpei/elc/history-of-elc.html (accessed on 25 April 2020).
39. Rohe, D.L.; Fall, R.P. A miniature Battery Powered CO2 Baited Light Trap for Mosquito Borne Encephalitis Surveillance. Available online: https://eurekamag.com/research/000/583/000583985.php (accessed on 7 March 2019).
40. Ritchie, S.A.; Cortis, G.; Paton, C.; Townsend, M.; Shroyer, D.; Zborowski, P.; Hall-Mendelin, S.; Van Den Hurk, A.F. A simple non-powered passive trap for the collection of mosquitoes for arbovirus surveillance. *J. Med. Entomol.* **2013**, *50*, 185–194. [CrossRef]
41. Meyer, D.B.; Johnson, B.J.; Fall, K.; Buhagiar, T.S.; Townsend, M.; Ritchie, S.A. Development, Optimization, and Field Evaluation of the Novel Collapsible Passive Trap for Collection of Mosquitoes. *J. Med. Entomol.* **2018**, *55*, 706–710. [CrossRef]
42. Panella, N.A.; Crockett, R.J.K.; Biggerstaff, B.J.; Komar, N. The Centers for Disease Control and Prevention resting trap: A novel device for collecting resting mosquitoes. *J. Am. Mosq. Control Assoc.* **2011**, *27*, 323–325. [CrossRef]
43. Service, M. *Mosquito Ecology: Field Sampling Methods*, 2nd ed.; Springer eBook Collection; Springer: Dordrecht, The Netherland, 1993; ISBN 978-94-015-8113-4.
44. Maciel-de-Freitas, R.; Eiras, A.E.; Lourenço-de-Oliveira, R. Field evaluation of effectiveness of the BG-Sentinel, a new trap for capturing adult Aedes aegypti (Diptera: Culicidae). *Mem. Inst. Oswaldo Cruz* **2006**, *101*, 321–325. [CrossRef]
45. Eiras, A.E.; Buhagiar, T.S.; Ritchie, S.A. Development of the gravid Aedes trap for the capture of adult female container-exploiting mosquitoes (Diptera: Culicidae). *J. Med. Entomol.* **2014**, *51*, 200–209. [CrossRef]

46. Mackay, A.J.; Amador, M.; Barrera, R. An improved autocidal gravid ovitrap for the control and surveillance of Aedes aegypti. *Parasit. Vectors* **2013**, *6*, 225. [CrossRef] [PubMed]
47. Reiter, P. A revised version of the CDC Gravid Mosquito Trap. *J. Am. Mosq. Control Assoc.* **1987**, *3*, 325–327. [PubMed]
48. Vazquez-Prokopec, G.M.; Galvin, W.A.; Kelly, R.; Kitron, U. A new, cost-effective, battery-powered aspirator for adult mosquito collections. *J. Med. Entomol.* **2009**, *46*, 1256–1259. [CrossRef] [PubMed]
49. Kompas, T.; Chu, L.; Pham, V.H.; Spring, D. Budgeting and portfolio allocation for biosecurity measures—Kompas. *Aust. J. Agric. Resour. Econ.* **2019**, *63*, 412–438. [CrossRef]
50. Kompas, T.; Che, T.N.; Ha, P.V.; Chu, L. Cost–Benefit Analysis for Biosecurity Decisions. In *Invasive Species: Risk Assessment and Management*; Cambridge University Press: Cambridge, UK, 2017.

MDPI
St. Alban-Anlage 66
4052 Basel
Switzerland
Tel. +41 61 683 77 34
Fax +41 61 302 89 18
www.mdpi.com

Tropical Medicine and Infectious Disease Editorial Office
E-mail: tropicalmed@mdpi.com
www.mdpi.com/journal/tropicalmed

www.ingramcontent.com/pod-product-compliance
Lightning Source LLC
LaVergne TN
LVHW070658170726
843515LV00001B/112
9783039433483